C000192818

ONLY A GOD CAN SAVE US

HEIDEGGER, POETIC IMAGINATION AND THE MODERN MALAISE

HENK J. VAN LEEUWEN

ONLY A GOD CAN SAVE US

HEIDEGGER, POETIC IMAGINATION AND THE MODERN MALAISE

HENK J. VAN LEEUWEN

Common Ground

First published in Australia in 2009
by Common Ground Publishing Pty Ltd
at The Humanities
a series imprint of TheUniversityPress.com

Copyright © Henk J. van Leeuwen 2009.

All rights reserved. Apart from fair dealing for the purposes of study, research, criticism or review as permitted under the Copyright Act, no part of this book may be reproduced by any process without written permission from the publisher.

The National Library of Australia Cataloguing-in-Publication data:

Only a god can save us: Heidegger, poetic imagination and the modern malaise
Henk J. van Leeuwen

Bibliography.
978 1 86335 631 2 (pbk.)
978 1 86335 632 9 (pdf)

1. Heidegger, Martin, 1889–1976.
2. Philosophy of nature.
3. Environmental ethics—Philosophy.
4. Human ecology—Philosophy.
5. Nature—Effect of human beings on--Moral and ethical aspects.

113.8

Cover image by photographer John Willems

For Kathleen and Paul

Contents

Introduction

"...if the individual is not truly regenerated in spirit, society cannot be either as it is made up of individuals in need of redemption"

(Carl Jung, *The Undiscovered Self*)

The world today seems to be in the grip of a malaise, where a growing feeling of emptiness, meaninglessness and vulnerability is entangled in a synergic relationship with the global ecological crisis. Hallmarks of contemporary life, such as consumerism, utilitarianism and market economics are increasingly gnawing at the very sense of what it means to be a human being. The World Health Organisation has predicted that by 2020 depression will be the second leading contributor to the global burden of disease. In affluent countries alcohol and other drug abuse is an escalating problem, particularly among the young. Ecological, political, social and economic worries, weigh down heavily upon the modern melancholy. Reinforced by the global financial crisis gloomy worldviews are becoming pervasive, whereby not only a sense of relevance and significance, but also the planet's continued habitability seems to be slipping from our grasp. As discourse on the environment reaches unprecedented levels, both in quantity and intensity, there is an awareness that things have got serious. Now that the negative consequences of human activity impact on the great systems of the biosphere itself, we are faced with a crisis that that can no longer be ignored. Global climate change seems inevitable with its wide-ranging consequences and its assurance of the further diminution of earthly diversity, essential to the health and beauty of the planet. Today, there is no guarantee that the West's 'living standards' can be maintained; even the continued survival of the human race is under threat if we carry on business as usual.

The modern idea of being in control and able to fix anything has come back to bite us. Yet the familiar response is to push nature further; to build more and bigger dams, to pipe the carbon back under the ground, to genetically manipulate organic life to better suit our wants and perceptions of the ideal, to imagine we can harness and contain nuclear processes that have their being deep inside the sun without paying the price. The smug illusion of human self-importance belies the miniscule human presence in the earth's 4½ billion-year history, where the sorry state of the planet now demonstrates that the earth can do quite nicely without us. Science and technology have provided the means to completely destroy humankind in a few hours. Yet, as Alan Weisman demonstrates in his book *The World Without Us*, if we were to vanish suddenly, nature would rapidly (in cosmological time) reorganise and repair itself into complex, diverse and well-functioning ecosystems, leaving only the crumbling ruins of once dominant civilisations. Yet, human activities continue on the premise of possession and mastery of nature.

The everyday conception of the natural world is predominantly through vague feelings and desires. Perceived immediate personal needs and wants are to be satisfied before considering what seems to be secondary; the needs of what is 'out there', that which we vaguely name 'the environment'. The prevailing way of thinking about nature is whether it is convenient (e.g. the weather); does it make us comfortable and safe? Is it productive, entertaining and attractive? Does it live up to our expectations or is it letting us down? Now, a more urgent concern has entered this thinking. Nature is losing its ability to absorb the consequences of our actions to such a degree that we may not be able to rely on it to provide for human needs and desires in the future. The global environmental crisis, now brought into clear focus, has made the *relationship itself*, the interaction between human beings and nature, a matter of concern. The contemporary culture's basic association with its environment is driven by economic theory and utilitarianism, leading to a disregard and estrangement from the natural world. Reducing nature to little more than a collection of ecosystems, to be valued and traded as capital assets or resources, goods and services, clearly demonstrates a shallow human relationship with the earth. By and large, the social relationship between humans and nature is considered via the various domains of sociology, eco-psychology and environmental management. However useful these may be, they tend to address the subjective positions of objectified persons and do not go the core of the fundamental nature of the human association with the natural world. Here I propose that attitudes, perspectives and actions that lead to social and environmental problems are all indicative of an absence of authentically *'being at home'* in the world. This buried underlying problematic association between humanity and the earth must be attended to if we are to make this relationship genuinely sustainable.

I suggest that the ecological crisis is inexorably interconnected with a deeper concern, sometimes articulated as "a crisis in the human soul"; a metaphysical crisis where a sense of meaninglessness threatens and darkens the human spirit, cultivating the roots of societal and ecological problems. At its source lurks an existential predicament where nihilism denies the meaning-fullness of life. The deficient relationship between humans and the natural world arises out of the *loss* of a sense of an essential earthly ground out of which humans derive significance and a personal sense of wellbeing. Such a primary foundational perception about the place and fundamental nature of human beings affects all other perceptions and therefore has far-reaching implications.

The modern worldview likes to imagine that it is in charge. Yet an underlying and growing sense persists that there is something vaguely disconcerting which is ultimately beyond our grasp and continues to elude us. This unease itself is not new. It is not something that has only appeared alongside the seemingly endless possibilities and information that modern science and technology has provided, albeit unequally, for the human race. It has not dawned in the growing affluence of the middle classes. Depending on the culture, tradition and historical circumstances, this anxiety may change in the way and the degree it manifests itself, but not in its deepest essence. This dimly perceived disquiet is the existential angst that is a permanent mark of the being of human beings. Pre-industrial and so-called primitive societies may well be largely free of the stresses of an overabundance of modern choice, but not of this primordial angst. At the same time, we are unlikely to want to exchange modern freedoms for the terror of the unknown, the perils and unpredictability of fickle gods and nature, and the grim and constant struggle for physical survival. Nor do we want to deride the amazing achievements and benefits of science and technology and advocate a return to some archaic way of life. Yet, although in Western societies we may be, by and large, exempt from the kinds of anxieties our forebears faced, an underlying uncertainty and unease about the human place and role in the world persists and indeed appears to be intensifying. This suggests that there is something happening here, the origin of which is not so easily pinned down. Perhaps in a modern society, as worries about everyday physical survival and well-being retreat, and the dominant techno-scientific perspective separates us ever further from our habitat, the real nature of the underlying angst forces itself upon us more directly. This deep unsettling anxiety then festers into a poverty of spirit that, I believe, hovers in the background of much of the world's ills.

So, the fundamental anxiety is not caused, as is sometimes suggested, by too much choice: the stress of having to decide what to buy, what to wear, where to have one's holidays, when and how to go, etc. This is merely the contrived freedom and hassle of supermarket choice. It may trigger a certain degree of stress, uncertainty and indecision, but is not what causes a deeper sense of alienation, unconnectedness, of meaninglessness. The urgency and importance we apply to consumer and recreational choices may well be symptomatic of this deeper angst. Too much information is another often-stated source of modern stress. Yet, does this not give us a hint that to overcome contemporary predicaments we need more than 'cleverness' and knowledge? The dissociation of the pursuit of knowledge from more fundamental concerns of how we are to live is, in itself, irrational; it cannot lead to a genuinely sustainable way of being. I suggest that this hint is a warning signal that calls for the retrieval of mostly forgotten or hidden insights, essential for grounding the pathway of life in the modern world.

If we are to better understand what it means to be human we need to come to terms with the enduring and pressing angst that seems to have its origin in the nothingness of human finitude. This existential void is a fundamental trait of the human abode, resulting in the uncanny kind of homelessness that I explore in this book. The concern with the meaning and origin of existence as such, alludes to something that is unique and authentic about the being that is human. For philosophy, this unescapable feature of both our being and our not-being persists as an inexhaustible and alluring source for its questioning. Of course, religious traditions have attempted to mitigate the existential angst via the concept of God as 'the highest being', the source and cause of all that is, and wherein resides the dream of life beyond the grave. As Sigmund Freud postulated, religion may be viewed as an attempt to come to terms with humanity's 'transcendental homelessness'. In theology, questions around the meaning of life and the foundation of humanity are usually elaborated in terms of our relationship with this God as a supernatural being, whose special status of not requiring further explanation itself continues to present an intractable problem. For many, a loss of faith has become a growing and powerful factor in a modern malaise that sharpens the sense of earthly homelessness which a simple appeal for a return to faith cannot address.

The existentialist thinker Frederick Nietzsche, some 120 years ago (1882), may well be marked as the first modern philosopher to jolt us out of complacent representational images of God with his articulation of the 'death of God'. His shocking declaration was meant to shake us out of our slumbers and wake us to a state of affairs where we can no longer remain the 'lazy slave' who relies on a, hopefully, benevolent master to look after us, give us the rules of life and to administer justice accordingly. For Nietzsche it was time to seek the proper role of human beings: to seek what it means to be, 'to become what we *are*', without the slave morality. As the meaning of this pronouncement is generally barely understood, it has been doggedly rejected by those who continue to seek solutions and solace in doctrinaire religious belief to the problem of personal purpose and temporality. In modern times, others have softened patriarchal and totalitarian images of God, yet questioning of the foundation from which their responses originate tends to be vague and haphazard.

Despite, or perhaps because of, declarations of a recent resurrection of God, particularly in assertive versions of the Islamic and Christian traditions, it seems that strident atheism is now also on the move. God's demise is again on the agenda. Authors and academics such asRichard Dawkins, Michel Onfray, Christopher Hitchens, Daniel Dennett and A. C. Grayling have again declared war on 'unscientific, irrational and unconditional faith'. While religious belief continues to provide a cognitive, supportive, moral and aesthetic framework for many,the success of books that are critical of religion and faith also reveal a growing scepticism about the tenets of established religions. The arguments of Dawkins *et al* ring a bell for those for whom the idea of an ultimate creator and controller of everything no longer makes sense. The God of metaphysics appears increasingly superfluous when simply used as a 'God of the (ever-shrinking) scientific gaps'. Today it seems ever more absurd to believe in a deity that is outside the laws of nature and takes a personal interest in even the most trivial of our daily affairs, yet does not seem to extend this to the life and death needs of billions living in abject poverty and under the constant threat of violence.

However, it leaves us to question whether it is still possible in the modern world, a world where everything seems explainable, do-able, measurable and controllable, to experience awe and wonder about life itself. The existential void is inevitably left exposed more chillingly in the wake of God's demise. We may ask whether the above books offer a credible and adequate (to a sense of human worth) alternative vision, or do they simply contribute to a growing and inevitable sense of disenchantment?Relinquishing God raises the question as to where that leaves us as human beings. Can life still be meaningful, or are we condemned to flounder in a nihilistic desert devoid of signposts? From where will we get our values; what will guide life's journey? What is the place of death in this journey? Does anything really matter, when science seems to be able to describe and explain everything, even the ambiguous workings of the human brain? Moreover, do we need to unconditionally embrace a humanist view that meaning (as distinct from *purpose*) in human existence has its origins solely in human beings?

Clearly, as the above authors confirm in their books, ethics and values are not dependent on the rules of religion; they arise from something deeper and prior to these traditions. Yet, I suggest, the recent flurry of anti-religion volumes do not adequately attend to a coherent foundation that authentically grounds human beings. They do not address the enduring yearning for a deeper sense of meaning; a hunger which in itself is not to be dismissed as meaningless, as it is part of who we are. This book grapples with the perceived emptiness and hones in more specifically to its corresponding angst. It draws us back to the existential void that has always been with us, but has been covered over by the consolations of religious conviction or obscured by the smouldering ashes left behind in strident polemical anti-religious assaults. While this void precipitates the threat of nihilism, which denies the existence of truth, goodness and wholeness, this book shows that it also holds a glimpse ofthe promise of an inexhaustible source of deepsignificance. The emptiness left in the wake of God's 'death' is not a nihilistic nothingness, but an opening that allows something new to emerge. This strange foundation may well give us the basis from which to build, not merely instrumental policies and strategies, but a wisdom that is able to address contemporary predicaments in a way that reveres the enduring relationship of human beings with the totality of nature.

Perhaps out of the ecological crisis nature cries out at us, but it does so from its own terms and concerns, within an order of balance and reciprocity in an unfamiliar manner that is not so straightforward. To genuinely respond to this call we need to be sensitive to its non-instrumental meanings, hints and gestures, which requires a more poetic imagination that is able to grasp our fundamental nature as conscious earthly dwellers. Contrary to the current dominant utilitarian approach, an ethic of environmental accountability cannot simply be prescribed; it needs to arise from something more fundamental that does not disregard the human soul. A deep-seated change in the human-nature relationship cannot take place by simply reaffirming 'values', objectives and priorities, or by using more efficient methodologies and technologies, but involves a greater *attentiveness* to what it means to be a human being.

This book aims to tackle the existential homesickness, which in its modern setting has spawned its own unique fears, responses and possibilities. Earthly homecoming that is true to the human essence involves meditation on the strange realm of the void; to seek a ground that is always near but has been covered over, forgotten or run away from. This quest is of critical importance if a new thinking is to emerge that understands contemporary dilemmas and charts a genuinely human future through a regeneration of the human psyche. I believe that at its most fundamental level, Martin Heidegger's later philosophy may well a basis from which we can ensure genuine enduring environmental sustainability. An authentic deep concern and reverence for the natural world can only come from a transformed conception of how we see our place and role within the earthly relationship. This conception needs to be rooted in an ontological soil in which the experience of existence, of 'being as such', can grow.

We will see that for the restoration of a genuine relationship with the earth we need 'gods'. As Martin Heidegger famously said in 1966 in an interview with *Der Spiegel*, "Only a god can save us"[1]. Although this remark sounds unhelpfully fatalistic it has nothing to do with the unconditionality of faith; it does not mean waiting for divine intervention. To retrieve a place for gods in modern life is not a summonsing for a return to religiosity. The 'god' or the godly I shall explore has little to do with any kind of being, and certainly not a 'highest being'. It does not aim to reinstall the Gods of metaphysics to their role as objects of worship or as the personalised supportive framework for human frailty, insecurity and uncertainty. We will

---

1.The interview is available on http://web.ics.purdue.edu/-other1/Heidegger%20Der%20Spiegel.pdf.

discover that the gods are implicated with events or moments of deep significance in the experience of human life; occurrences that shed light on a way forwards that is faithful to the human essence.

Thereby I explore something that is utterly original in his later work and overlooked in much of the work carried out on 'Heidegger and environmental philosophy'. Moreover, this salvages core insights, lost in the current fashionable rush in some philosophical circles to demote his thought. Here we seek what Heidegger's journey into the sphere of the question and the meaning and truth of Being actually *yields* in a quest for the retrieval of significance of being human in the world, i.e. for earthly dwelling. It offers the possibility to genuinely ground an authentically human response to contemporary problems. Therefore, my approach is to take up his insights in a way that sheds light on these problems, rather than this being another treatise on a particular Heideggerian theme.

Unless the individual spirit is regenerated, the collective spirit of society cannot be healed. Here I seek a fresh sense of the significance of what it means to be human without recourse to religion, mysticism or various forms of vague thinking. However, as we shall see, this does not preclude an unfamiliar kind of spirituality; it even expressly seeks it. This book is for those who are sceptical of, or have rejected, religious faith, yet question whether a purely instrumental or utilitarian view of life is adequate to express the full potential and mystery of human experience. It has one thing in common with religion: an exclusively rational, materialistic and instrumental view of life does not tell the whole story of what it means to be human. The intuitive sense and need of 'something more persists as a genuine feature of the human condition. However, it does not accept religion's central precept of a miraculous realm, ruled over by a supernatural entity who is able to suspend the laws of nature.

In the wake of the existential and the ecological crisis humanity finds itself at a crossroad. This not simply an intersection involving two choices: one where we continue as we have and thus destroy our habitat; the other where we attempt to somehow ensure our survival via the tools of science and technology. I suggest that the chances of success with the latter are by no means guaranteed. Even if we do survive, the integrity of the earth and our quality of life are likely to be much diminished. The current global financial crisis is forcing change upon the overindulgent culture. Now to find the means to change the way we live calls for a paradigm shift from habitual thinking. Perhaps such a shift would in fact move human beings closer to their true dwelling place. I believe that if we pause at the crossroad another such way may be distinguished. This is not a highway with its boundaries assured, its signposts clear and unambiguous and its surface smooth and unencumbered. It is instead a modest pathway that must be trodden carefully; it cannot be rushed without stumbling over its difficulties or missing the sights along the way. It is necessary to stop and attend to these sights as they are the insights that illuminate the pathway itself that leads to, what we might provisionally call, 'wisdom' [2] . This is the wisdom needed to re-vision the human place in this world in a way that is faithful to our selves and to the earth.

"Does not the sacred lie within what is already meaningful for us...

on the other side of the already-known?

And don't we have to be our own guides in seeking out this other side, since only we can identify 'the known' in our lives?"[3]

To embark on the venture of this pathway requires a fissure in the every-day; a radical break in the familiar and habitual thinking of the "already-known". To find a 'self', we need to depart and wander.

2.Single quotation marks or scare-quotes may express irony and qualify the use of a word or phrase. Sometimes their use may be for specialised, idiosyncratic uses of everyday expressions. In a book such as this, often dealing with indefinable and un-familiar concepts, it seems that they are often called for. This is particularly the case when I refer to meanings that are other than the usual correspondence. At times, single quotes exclude a first sense and focus attention on another less tan-gible one. Or it may serve to distance the reader from a too naive understanding of the quoted expression. Single quotes may be regarded as a kind of meaning-filter. They may single out one particular connotation, while screening out others, e.g. as 'in the sense that I have introduced', or 'in a certain sense'. Double quotation marks refer to quotations from the literature. The use of italics is mainly one of emphasis.

3.*Freya Mathews. Journey to the source of Merri Creek, Melbourne: Ginninderra Press, 2003.*

Part 1

Preparations

Chapter I
Herr Professor Heidegger

Much has been written about Martin Heidegger. The temporal and philosophical emphases and shifts in his pathway of thinking and his astonishing output over some 65 years are a rich source for academic debate. Controversies have centred, not only around his philosophy, but also about the man himself. His thought and life have been subjected to many interpretations and responses, ranging from naïve and selective approval, to vehement contempt of his work and of Heidegger as a person. Both extremes bar the way of constructive engagement with his thinking and make it impossible to carry it further in a practical way. I strongly believe that his philosophy holds insights that are deeply significant and relevant for contemporary quandaries and that provide fertile ground for further exploration. Yet, as this is not a book *about* Heidegger as such (there is no shortage of such volumes), I do not intend to delve too much into the controversies surrounding him, other than to comment briefly on some common criticisms and issues in order to clarify the approach that I take here.

So what kind of a man was he? It has to be said that one is unlikely to find much humour in Heidegger's writing. Thinking of Being as such is for him a serious matter and there is little doubt that he regards himself as foremost amongst such thinkers. Occasionally, such as in his asides regarding the thoughtless contemporary state of affairs or perceived misunderstandings of his texts, one encounters his corrosively acid sarcasm. In the glimpses afforded to us, the man himself at a personal level appears deeply

conservative.[4] Although I take a divergent approach, his views on technology and instrumental science are often interpreted as being anti-modern-
ist. He is polite in a formal rather impersonal sort of a way. While he seems
to lack warmth, and over the years clearly demonstrates that he is not exempt from universal human weaknesses, he does not appear in any way
contemptible. In a prevalent rush to demonise him as some kind of
Gestapo henchman his alleged 'Nazi-past' is often used to paint him as a
rather despicable person, whose philosophical writings must be approached with the utmost care, lest one is infected by the underlying, insi-
dious virus of National Socialism. It should not be forgotten that Hitler in
the early 1930s was widely admired as a great national statesman, both in
Germany and abroad. Heidegger's involvement with the movement resulted in his short-lived appointment as rector of the University of Freiburg in
1933. Undeniably, at that time he perceived something in National Socialism that would counter the malaise of the times. Like Nietzsche before
him, he was deeply worried that scientific-technological discourse was
spreading to all realms of life. Thereby he believed that National Socialism
possessed the necessary authority able to reverse the increasing instrumental and objectifying domination of the empirical sciences in the German university. He considered that these were appropriating any reflection
on the ontological and metaphysical *foundations* of scientific presumptions
and certainties. Yet, I regard his involvement as transitory whereby he
soon returned to his life-long project of questioning the meaning and the
"truth of Being". Much of his developing thinking following this period
may be seen as grappling with the errors of his earlier dalliance. Perhaps his
astonishing output in the late thirties was his attempt to rectify and pre-
vent misinterpretation of his thought.

Nevertheless, around that time Heidegger seems specific about his vision for a spiritual progress of the German *Volk* as historical grounding, heralding another beginning as a new epoch in Western history. It is a dream
for a "people of poetry and of thought".[5] This dream for a new relation to
language and thought is also expressed in a passage by Wilhelm von Hum-
boldt, which Heidegger in the essay '*The Way to Language*' quotes approv-
ingly: "an increased capacity for sustained thinking, and a more penetrating
sensibility, ...introducing something new into language ...which would be a
lasting fruit of a people's literature, especially poetry and philosophy"[6]. I do
not believe that this can be interpreted as a notion of a triumphant ascent

---

4.See e.g. the communications published as the *Zollikon Seminars* (ZS). His long
friendship with Swiss psychiatrist Medard Boss resulted in an extensive dialogue
(recorded in ZS) where he used his pedagogical skills with psychiatrists and psycho-
therapists, untrained in philosophy. Also, see Adam Sharr, *Heidegger's Hut* (MIT
Press, 2006) and *Martin Heidegger Letters to his wife: 1915-1970* (Wiley, 2008) for further intriguing insights into the life of a fallible man.

5.EHP p.48.

6.OWL p.136

of German civilisation. Heidegger at the time was reminded that the fleeting heroes of the era are always surpassed by that which transcends us all. The problems of his hope for an onto-political realisation of this philosophical objective will continue to be debated, contributing further to an already burgeoning body of literature. However, analysis of his writing too often place unnecessary emphasis on a historical (as chronological history) and on the concrete everyday manifestation of his dream, rather than something thoughtful that occurs outside sequential historical events. Heidegger shared Nietzsche's conception of time as something to be grasped as a whole, instead of its dominant sequential notion. The translators of Heidegger's *Mindfulness* draw attention to the error in the attempts in historicising his thought[7], yet this is what some scholars continue to engage in. These interpretations thereby bury the possibilities of the seeds of his thinking so deeply, that their concealed, yet exquisite, potential is smothered.

Except for these remarks, it is not my purpose to join in this largely unproductive contest. When first beginning to study Heidegger, I, like many who at first find his locutions difficult, turned to the plentiful commentaries and critiques of his works. Although these clarified certain issues, I almost came to a point where it seemed that relinquishing his writings in favour of other, less 'tainted' thinkers, might not be a bad idea. Then I decided to simply return to the works themselves; to unhurriedly encounter his *ways* of thinking. This, I suggest, is how his writings should be approached: *simply as we find them, and as they address us, without any predetermined ideas.*

Heidegger himself often insists that his are not *works* that define a world-view, but rather are *ways* (paths of thinking and questioning) that are neither constrained nor pre-determined by epistemological and cultural baggage. The guiding motto he places at the beginning of his vast "Collected Works" (*Gesamtausgabe*) is "ways, not works". By this he emphasises that these are not to be seen as the works of an author, whose 'opinions' stand at the centre of attention, to be the subject of scholarly debate and dissection. His desire was rather that such writings seen as *path ways of thinking* (*Denkwege*), to be engaged on their merit. Already at the conclusion of his early Magnus Opus, *Being and Time*, he comments: "One must look for *a way* to illuminate the fundamental ontological question and *follow* it. Whether this is the only way or even the right one at all, can only be decided after we have *followed* it. The strife in relation to the question of Being cannot be settled because it has not even been started"[8]. *Being and Time* was a preparation for the venture that was to take him into unchartered realms in the mid 1930s. This venture that follows an uncertain and difficult pathway is not just for Heidegger but may be taken up by anyone who is not content to be merely comfortable and wishes to seek what

---

7.M p.xi. *Mindfulness* was written in 1938/9.

8.BT p.398.

lies beyond the familiar. The way such a venture pans out will be as unique as every individual who seeks the source of that which seems to call us to questioning. As Heidegger said in a very late work, there are times to: "sow a seed here and there, a seed of thinking which some time or other may bloom in *its own way* and bring forth fruit"[9].

If we approach his writings with an open mind there is little in Heidegger's reflections that resemble the rhetoric, and even less the enactment, of the movement of National Socialism. What is more, his admonitions on the need of stillness - in 'listening' as well as 'deciding', his rebuke of art as cultural achievement and symbol of superiority, the need to awaken to the modern danger of, what he calls, "calculation", "acceleration", "machination" and the "outbreak of massiveness", all these are as utterly counter to the prevailing political and military situation of the time as they are today! The homeland he dreams of in his essay on Hölderlin's poem "Remembrance" is the place where the poetic spirit is at home. Although as a German thinker he hopes to renew a German poetic spirit, the vision could be applied anywhere where there is both 'a sense of place' and a sense of 'nearness to the source'; near to one's essence and one's dwelling ground. However some scholars continue to make forced associations, whereas less distorting and more positive connections could just as easily be made. I suggest that the time has come to encounter Heidegger's writings on earthly 'homecoming' as they come to meet us, and speak to us as open sites for the unfolding of a deeper truth, without the shadow of National Socialism darkening what is being illuminated. In his *Elucidations of Hölderlin's Poetry* Heidegger writes: "Homecoming meditates ...on the Holy"[10]; it seeks what is already near but mostly forgotten. Heidegger here is seeking a different truth from everyday rationality; a transformation in thinking that is wholly at odds with the common sense of the day. "The treasure" that he thereby approaches, one which "is most proper to the German homeland", is the same treasure to be found in the "joyful origin" of all earthly dwelling[11]. Being near to this origin, a realm explored in this book, preserves a ground that is most appropriate to the people of *all* homelands.

Central to the book is the *'crossing over'* from an instrumental, possessing, controlling and self-important attitude to another, more authentically human, way of thinking. This is not a pathway to be construed as some kind of mass movement. It rather involves a holistic awakening and an essentially personal mindfulness; one that is nevertheless not isolated, as it is bound to the totality and deep significance of all earthly beings. The crossing ventured here is deeply implicated in, what is called, Heidegger's 'turning' in thinking about 'Being as such', taking place in the mid 1930s. Being is the source that surpasses human beings, which over the history of humankind has been given many metaphysical names, yet for Heidegger is

9. ID p.74.

10. EHP p.222.

11. EHP p.43.

un-nameable, un-representable and essentially hidden[12]. Following the more analytical approach of *Being and Time* he became aware that if modern humans are to grasp the way 'Being as such' has been forgotten, we need to retrieve a different, essentially poetic, language that is able to express the truth of Being. This is a turning to a way of being in which a proper relationship with the world and with our very selfhood, as mortal earthly dwellers, is regained. Heidegger regularly lamented that his turning in thinking is often misunderstood: namely that he is presenting some new theory or system of truth rather than reflecting on the need for questioning the unfolding of Truth itself. "The epoch of philosophical 'systems' is over" he proclaims rather ostentatiously[13], as he stresses the need for another way of questioning. This other thinking is a kind of deep reflection. The sense of this meditative thought I shall develop later, as it is inexorably implicated in a fresh perception of human beings as authentic earthly inhabitants.

As it is not simply objective or subjective, the other thinking is not easily brought to articulated language, and requires a pathway, or a crossing, where one gradually grasps its significance as one's thinking is transformed. Heidegger, referring to his own writings, declares: "...whoever then compares and reckons the interpretation with the already existing views or exploits the interpretation in order to 'correct' the existing views, *he has not grasped anything at all*"[14]. Again, this is a plea for simply *encountering* his thought, which in itself is an encounter with something fundamental or primordial; something he believes has been unquestioned since early Greek thinking, yet is at the core of Western philosophy. His approach is particularly pertinent here, as my aim is to also engage the specific pathway of the 'in-between' of Heidegger's crossing in an accessible yet thoughtful manner. To accomplish this requires an unfamiliar venturesome approach; one that is not too polemic and abides reflection. As Heidegger writes, "But thoughtful questioning is not the intrusive and rash curiosity of the search for explanations; it is the tolerating and sustaining of the unexplainable as such"[15]. Astonishment is the outcome of such transformed thinking: "Astonishment includes a decisive suspension of position-taking"[16]. Yet, such position-taking is demanded by many scholars. It is true that philosophy usually is an argumentative discipline that thrives on disagreement.

---

12. There is ongoing controversy about the capitalisation of 'being'. Following Michel Haar (*Heidegger and the Essence of Man*, trans. William McNeill, Albany: State University of New York Press, 1993), I capitalise 'Being' when referring to its most primary and elemental sense, i.e. in the manner of the grand metaphysical names in the history of human thought such as Truth, Substance, Subject, or Will. When written with the small 'b' I emphasise the active sense (*to be*: understanding it as a non-substantive verb) of the unfolding of the mystery of how things come to be.

13. BQP p.125.

14. M p.372. Note also his sarcastic comments on "the gang of the curious" in M, p.378.

15. BQP p.148.

16. BQP p.143.

However, to seek parallelisms between different thinkers and seek what is truly fundamental in their thinking helps us to move forwards and perhaps uncover something unique and of genuine benefit.

Here we seek, not 'the meaning of life' as an explanation of the world and all its details, but rather *what it is that makes life meaningful*. The re-visioning of perceiving the human place on the earth is quietly radical. It is a journey of homecoming to one's essential place. Earthly dwelling cannot occur without remembering one's dwelling ground. We will see that the veiled manner in which humans belong to this fathomless inexhaustible origin is 'strange', i.e. unfamiliar and hard to identify. Yet, realising the *need* for the retrieval of a ground of significance for human earthly inhabitation breaks the cycle of the contemporary impoverishment of the human essence and transforms the shallow human relationship with the earth. If there a consequential relationship with the totality of nature, the 'oughts' of environmental ethics become a desire for an earthly way of dwelling that does not wear out the earthliness of earth. We will see that for the restoration of the ground of human dwelling we need 'gods'. And to retrieve these gods that 'save' in daily lives, modern humans must learn to risk the unfamiliarity of the realm that is beyond the habitual and the already known.

The famously ambiguous remark, "Only a god can save us", was uttered by the 77-year old Heidegger in an interview with *Der Spiegel* on the 23rd of September 1966. In the first part of this interview he was questioned about his involvement with National Socialism around 1933 and the political developments thereafter. The adequacy of his responses is still a matter of controversy, but need not concern us here. His statement that to solve the dire problems facing the world requires the return of 'gods', has overshadowed his legacy since his death in 1976. There have been various more or less helpful interpretations of this pronouncement, with perhaps Richard Wolin's judgement of a "sad profession of impotence" being foremost in the latter. Anyone who has read Heidegger a little more thoughtfully will understand his comment does not mean waiting for divine intervention. It is not a kind of fatalism, but rather a thinking that goes beyond activity and passivity. He stresses that human beings are being uprooted in their essence and the only thing that can save them is the gentle power of 'thinking and poetry'. Such *poetic imagination* mediates the 'gods', which are the moments of deep significance, which have within them the possibility of directing us to genuine human homeliness. These help to get us ready for a change in the self-destructive ways we deal with the world.

Heidegger is realistic when asked how this thinking may be put into practice: "I know nothing about how this thinking is 'effective'", he responds. He regards the roles of public commentator, of preacher and moral judge as being contrary to his allotted task. He unassumingly adds that perhaps if it is understood as a pathway of thinking that leads toward an inner silence, it may yet endure for a transformed way of being. The interviewer's frustration can be sensed in the latter parts of the interview: why is Heidegger not prescriptive; why does he not give us some clearer ideas as to how we are to live; what we are to do? Heidegger's response is infuriating for journalists who expect instant solutions to deep-seated problems. Thoughtful reflection that allows us the necessary time and space to understand the essence of contemporary predicaments does not figure in their tight timelines. What the interviewer does not come to terms with is that Heidegger is first and foremost a *thinker*. A thinker's job is to think. He or she then gives us a hint as where to look; whether we respond and moreover, whether we will *see*, is up to us. Therefore, it is up to *others* to grasp the treasures he uncovers in his pathways of thinking. This book takes up the challenge to apply their potential for the emergence of a fresh and more meaningful approach to the human relationship with the earth. Poets and thinkers such as Heidegger cannot exhort their audience into action; yet they can inspire genuine transformation by touching the very essence of the human being in the world. All he claims is that the way of thinking he uncovers, could awaken, clarify and strengthen our readiness for a new era in the human sense of being. He truly regards his primary task as "finding a base for thinking itself". This is an assignment that he attempts via a somewhat monolithic inquiry into the meaning, truth and the experience of Being, which occupied him throughout his life. It is a task Heidegger takes very seriously and accomplishes with astonishing skill and care. He does not believe that this kind of thinking about what is most fundamental about existence can in any way be authoritative, in the sense of giving instructions. He is adverse to sound-bites and quick-fix 'solutions', which undeniably results in some difficult, at times obscure, writings. In the interview he simply pleads that we think about "the fundamental characteristics of the present age". Because the prevailing instrumental way of thinking makes it very difficult to do so, these are rarely recognised. It is to these "fundamental characteristics" of contemporary life that I now turn.

Saving us from what? The Modern Malaise

In affluent western societies there is a growing consensus that 'wealthier' does not necessarily mean 'happier'. We are assured by the gurus of statistics who measure and calculate everything that matters about us that rampant choice and rising prosperity (its un-sustainability now explicit and perhaps finally brought to account) has left our sense of wellbeing no better than 50 years ago. Despite most of us having all the necessities of life and more on call 24 hours a day, it appears that we always need 'something more'. As we chase the elusive mirage of pure bliss via the shifting fashions of consumer goods we find that as one dose of instant happiness fades, another one is called for in the form of more acquisitions, more novelty and diversionary amusements, but which in essence is more of the same. These inducements offer the ego only short term rewards on the treadmill to nowhere. The ego regards everything as centred on itself. Whilst this is not an entirely useless illusion (it helps us to function as persons), it can also lead to a wretched individualism which ignores that existence is always shared with the needs and 'being' of other beings, human and non-human. Moreover, it disregards the reality that life is a gift whose source is beyond our grasp.

[me]

Recently, I came across a major communications provider's glossy advertising brochure buried within my daily paper. Its gleaming cover was graced with a contrivance consisting simply of a 'square bracketed' [me]. Aimed at 'technologically savvy', 'lifestyle conscious' affluent and consumer-ready young, its message was all about 'look at me': my lifestyle, my 'friends', my 'entertainment', my 'image', my 'success', my unending 'youthfulness'. It was all centred around an enormous grotesque illusion: [me], immortal, beautiful, powerful, in-charge, irresistibly attractive and incredibly 'interesting'. Yet, all we really end up with is a mechanised being, a 'boxed-in' robot to be sure. Knowledgeable as far its access to software and electronic media allows it to be, reliable in a mechanistic kind of way, functionally conscious, but in fact also deeply boring and spiritually empty. As neurologist Susan Greenfield argues, the emphasis on process over content, or method over meaning becomes addictive, significantly altering a brain that is "exquisitely malleable"[17]. The spoon feeding via computer menu options rather than free-ranging inquiry, the decline in linguistic and visual imagination, the constricting and brutalising of language by text-messaging and 'twittering' that lack the verbs and conditional structures essential for complex thinking, all these contribute to the creation of a mindless generation marked by a total absorption in the here and now, and to the inability to reflect on past experience and future implications. She warns that "individuality could be obliterated in favour of a passive state, reacting to a flood of incoming sensations - a 'yuck' and 'wow' mentality characterised by a premium on momentary experience". The above kind of advertising reinforces this loss of brain connectivity and reflects a common attempt to negate what constitutes a human being by deceit and the manipulation of emotions, in order to ensnare its unwitting 'market potential'. The hype of commercial jargon is thereby also constructing, what Michael Leunig calls, a "fake...self-infatuated society", where illusionary self-importance, mass wellbeing and prosperity conceal the reality of an environmental and psychological catastrophe that gnaws at the psyche of today's generations[18].

Heidegger argues that the more 'important' we make ourselves out to be, and the more we polish inflated egos, the more our true essence is covered over and forgotten. This is the danger from which we must be 'saved'. He writes,

> *"Meanwhile man, precisely as the one so threatened, exalts himself to the posture of lord of the earth. In this way the impression comes to prevail that everything man encounters exists only insofar as it is his construct. This illusion gives rise in turn to one*

---

17. *The Age*, Good Weekend, June 14, 2008. As Norman Doidge also argues in *The Brain That Changes Itself*, Carlton: Scribe Publications, 2008.

18. Michael Leunig, *The Lot - In Words*, 'Love in the Milky Way', Penguin, 2008.

final delusion: It seems as though man everywhere and always encounters only himself.... In truth, however, nowhere does man today any longer encounter himself, i.e., his essence"[19].

The modern discourse is one, which ultimately means that:

If everything in the world, including nature and human nature, is a human construct birthed by the tools of utilitarian language and abstracted by instrumental preconceptions, then everything *seems* to depend on us. We look, we listen, but we only see and hear ourselves, as we have convinced ourselves that we are the only thing worth listening to. As a result, we neither understand nor experience, what it means to be truly human on this earth. It seems as though we have been *abandoned* by a meaningful sense of being.

Then, having lost a proper sense of the human essence, we no longer see ourselves as being any different from the things we attempt to put in order, categorise and manage. Heidegger calls this "calculative thinking", which does not refer to a practice of mathematics, but to a fundamental human comportment. The calculative and technological attitude seeks to order, quantify and classify. A characteristic of calculative thinking is its orientation towards, and reliance on, an overall goal of technical efficiency. As a *purposeful* activity, it *must* produce quantifiable outputs. In the creed of corporate planning, the human resource is expected to produce an output whose progress is calculated, measured, adjusted in terms of *what* is being done, *who* will do it and *what* will come out of it.

In this dominant utilitarian perspective, human beings, despite a prevailing sense of self-importance, are reduced to mere resources, ready for exploitation and disposable when no longer useful. In the mechanistic worldview it appears that we are little more than pawns of science and technology, or cogs in a gigantic mechanism. The rush to control and manipulate reverses to crush our creative imagination, whereby technical competence becomes the sole measure of human value and purpose. We are to submit to our technological destiny as our inevitable becoming. It is no longer enough just to keep up with technology; we are to be subsumed by it. As the growth of the technology/information juggernaut accelerates, our only option to avoid being crushed by it is to become 'faster' ourselves; to become 'more efficient', better planned, ordered, structured and skilled for our allotted role as consumer and resource. Then what it means to be human has contracted; the whole range of human experience and creative potential has shrunk into **[me]**.

Today we are told that we need to yield to 'the digital world'. About the greatest shame that can befall us is to be accused that we are "out of touch with technology"; we are mere primitives, belonging to an age long made redundant (by technology). It is ironic that such remarks demonstrate that thinking has been totally overwhelmed by the *essence* of technology, which has an unrecognised momentum of its own. It reveals an inability to think

---

19.QCT p.27.

in ways other than measuring, calculating, valuing, managing, manipulating, classifying and sorting. To be sure, the difficult human relationship with technology is sometimes vaguely lamented, resulting in various cautions and prescriptions to ensure 'balance'. However the question of why technology has its own unstoppable dynamic is not given attention. There is an illusion that technology, in its essence, is something that human beings have under their control. Yet, we are caught, claimed and challenged by the power of its essence in a way that makes us want to operate and manipulate everything in an instrumental way. To respond appropriately to this challenging does not mean we try to eliminate this fundamental human trait. A proper response rather begins with becoming *aware* of it.

In the instrumental attitude the quality of the experience of any activity is less important than the efficient (temporal and economic) achievement of tasks and the almost obsessive gathering of ever more information. This information is then accepted as *knowledge* to be packaged and implemented in hasty programs. What counts as knowledge is defined in statistical terms that smooth over diversity and plane down truth to a categorical sameness. Sometimes this knowledge is 'efficiently' distributed as mass information, which is not reflected upon, at times misguided and misunderstood, or worse, used as a tool to manipulate public opinion. Often information for 'public consumption' is to be merely 'interesting', as we observe for instance in many science or 'nature' TV programs. If information raises further questioning it is often centred on technical problems to be solved, rather than being a pointer to the possibility and need for real fundamental change in the face of today's challenges. The vague comment, "that's interesting!" is often one of indifference or of fleeting consequence. Or, there is the enthusiastic 'idle chatter' of those who find *everything* interesting in the world but nothing is ever deeply significant. In the fleeting experiences of manufactured events one *thinks* one has a goal and is 'living life to the full'. This may be temporarily pleasurable, exciting or even constructive, but are not goals of the kind that make human existence truly significant and fulfilling. Everything becomes a short-lived experience, to be tallied to one's inventory of experiences: "now that was interesting; but been there, seen this, done that; what else is there? What am I still missing out on?" What we are really missing out on is a sense of awe and wonder about life itself. Its absence causes a vague fundamental emptiness that confirms itself in the indifference of boredom or in the restless pursuit of assorted distractions, such as the flight into consumerism and the quest for instant amusement and good feelings. The lack of this sense underlies the existential feeling of meaninglessness that may intensify into psychological disorders and nihilism. It also lies beneath the environmental predicament, as it can only grasp a utilitarian earth, where nature, including human beings, is diminished to a mere object and resource.

Heidegger devoted much attention to the primordial pre-Socratic Greek thinkers who he interpreted as perceiving nature in a pre-metaphysical way that did not objectify everything. Theirs was a 'beholding', in awe and wonder, of Nature as such. This wonder is not as in its usual interpretation, where it soon becomes commonplace. It is not some augmentation or gratification in human life. It is not a state of amazement or fascination with something, but rather is more concerned with an awareness of the wonder of life per se; a genuine grasping of the emerging of concealed Nature in a manner that is rarely noticed today. Nor is it simple *curiosity*, which Heidegger in *Being and Time* argues is characterised by *not staying with what is nearest*. Of course, curiosity may well be the hallmark of mental freedom. It has delivered us a testable knowledge of the world and has contributed much to the understanding of our place in it. However, when it merely remains with consciousness it may become an impatient desire for the new and the exciting and thereby "makes present only in order to see and have seen". It does not seek to *grasp* what is objectively present by staying with it; to let it show itself in its essence and manner of appearing. Travelled once, walking the forest trail then no longer holds any allure. As soon as we catch sight of the "matter in question", curiosity or boredom is already looking for the next thing, forgetting of what went before. In *Being and Time* Heidegger writes that idle talk, curiosity and existential confusion or bewilderment are interrelated, and lead to the illusion of a wholly instrumental world. Then, guided by popular opinion and hearsay, as well as an underlying boredom, human beings are unable to discern between genuine understandings of what really matters, and what does not. This always gives to curiosity what it is looking for and to idle talk the illusion of having the final word, thereby leaving nothing of further interest. The language of idle talk omits going back to the *foundation* of what is being talked about, and thereby blocks any primary meaning. Paradoxically, it then *appears* that there is nothing that *requires* further fundamental unveiling.

Heidegger, in *Being and Time*, uses terms such as "idle talk" and "everydayness" to describe the customary and taken-for-granted, and essentially unreflective discourse of daily life. They are necessary parts of everyday life and therefore are not explicitly intended to be disparaging. They are not meant to express a disdain for the daily problems of morality, ethics and politics, as though they stand in the way of some inner way of pure being. It becomes a problem, of which we need to be aware, when there seems to be no other way of being, barring a primordially genuine relationship towards the world. To be attentive to the way the everyday works puts one in a better position to decide more thoughtfully for what is to come; a future that may then be revealed in a very different way from "the tried and the usual", or in the "chronicling of current happenings, whose future is never more than a prolongation of today's events" [20]. This Heideggerian aside is a reminder of a common state of journalism in the mainstream press today that feeds us 'all the information we need', processed and packaged for easy consumption. The familiar thereby has no chance to become the unusual; it has no possibility of again becoming

---

20. In 'Language in the Poem', OWL p.197.

something able to evoke a sense of awe and wonder. Moreover, the transitory experiences of the 'most unusual', such as in packaged entertainment, the events that amaze and excite, are soon consigned to the ordinary. Yet, the everyday way of thinking seems so obvious and self-assured that there appears to be no alternative. Then, drifting into existentially uprooted groundlessness, unable to encounter its fundamental essence, the human psyche is left dried up and fruitless.

So, what effect does this have on the modern relationship with 'the natural world'? In the everyday experience, nature, or 'the environment', is commonly seen as a problem or a commodity in need of 'management'. Or nature may be merely seen as a theme park; lakes, rivers and seas are just the thing for the thrills and spills of water-ski jetting, hills for the adrenalin-rush of trail-bike riding; spectacular scenery offers the photo opportunities of holidays. If nature is not 'interesting' or attractive, it is deemed to be wanting and in need of modification or augmentation. In the colonised world, nature has become a platform for the projection of human ambition and domination. When nature is exploited as resource, society acknowledges that its demands for materials, energy and transport all create wastes, yet, in the 'not in my backyard' or NIMBY syndrome, no community appears willing to absorb its share. The earth must be productive in a calculative sense; therefore farming and logging interests continue to demand compensation for *not* clearing land. In urban situations residents persist in throwing garden waste over the back fence into the neighbouring 'nature reserve', which they so often see as a scruffy bit of 'bush' and a fire risk anyhow and should be replaced with green lawns dotted with European trees. Often, about the only exposure urban dwellers have to 'the environment' is the weather, about which there is much conversation and complaining. We want water in our dams; gardens that are lush, but do not like the inconvenience of rain, especially when it interrupts the weekend barbeque! Clearly nature is often regarded as 'letting us down' and we wish we could control it more, as in the persistent desire, indeed even the *demand*, of making Australia drought or fire-proof, but not taking responsibility for the uncertain implications of interventional actions. Although in Australia traditional indigenous relations with 'nature' are often referred to as noble and commendable, they are not considered 'practical' or 'realistic', and therefore not taken up as a serious guide for a modern society.

The utilitarian and instrumental attitude drives most contemporary activities that impact on the environment. Mainstream discourse about the environmental issues assumes that as long as society is guided by rational scientific knowledge of natural systems, it can live in harmony with nature. The concept of sustainability is widely regarded as achievable, given sufficient technical resources and scientific 'know-how'. Sustainable development attempts to overcome the unacceptable obstacle to materialist modernity that environmental protection can only be achieved at the expense of human economic and social development. The term accommodates ideas of growth, change and development, i.e. expanding technological-economic systems, whilst aiming to maintain the long-term integrity of ecological systems. Despite continuing evidence of environmental deterioration and ecological impoverishment, it suggests that we can 'have it all'; ever continuing growth (now shown to be patently impossible) *and* a cleaner, healthier environment. In response to growing public perception that 'something should be done' we are witnessing an exponential growth in the corporate *rhetoric* of sustainability, while fundamental change lags far behind. In society at large 'green' talk is highly fashionable. Yet, the language and application of sustainable development has not resulted in recognition of necessary paradigm and lifestyle changes, but rather has re-embraced technocratic expertise to 'manage the environment', particularly through prediction and control with its underlying sense of the possession and utility of nature. However, the uniqueness and mystery of Nature as such is not apprehended, when the riches of *"what is natural... hardens into an increasingly dull mixture of prior possibilities..."*[21]

Human impact has clearly touched all corners of the earth. Even nature's great self-regulatory systems are now affected. This is a world that is overwhelmingly influenced by human activities and disturbed to varying degrees, from relatively pristine to places that are degraded so much that their natural features seem all but obliterated. Yet, although the sustaining and nourishing character of the earth may become seriously wounded, nature will nevertheless continue to self-organise and self-emerge from a 'wildness', generated from an innate 'essence' that is outside human control. Despite the illusion of mastery, human nature cannot exist independently of the world of nature. The social constructivist, Frank Fisher, argues that the activities of modern humans are inconsistent with the environment in which they are conducted. Models of nature cannot be applied to nature *as* nature without the flawed assumption that nature will always re-integrate itself around our incursion[22]. As nature has been abstracted into a collection of mechanical devices it has become a material manifestation of our theoretical models. While this is a convenient way to causally operate nature according to prevailing instrumental preconceptions, it diminishes nature to a 'human construct'. However, nature as the earth (as a whole system in the sense of Jim Lovelock's *Gaia*) is not a like a machine that can be controlled by any means and made our possession, but is more like a self-generating organism that is unpredictable, indeterminate, non-linear, and sensitive to changes within its environment. It is worthy of respect, not as a laudable opponent, but rather as that mysterious whole into which we find ourselves 'thrown'.

If the fundamental essence of being human is to be retrieved a deeper truth must be emancipated and safeguarded so that its 'saving possibilities' are preserved. Can we still hope for this truth to awaken wisdom? True hope is not ambition or empty wishing. The will of the ego constructs its own hubris, folly and delusion, and may actually displace hope in its

---

21.CP p.93.

22.Fisher, Frank. *Response Ability*, Elsternwick: Vista Publications, 2006, pp.25, 34.

attempt to fill up the vacuum left by the loss of this truth. Hope is rather a fundamental condition of the human soul that affirms the inexhaustible possibilities of being. However, this hope easily fades by the pervasive effects of what Heidegger calls *"machination"*. This is the totalising hold of technological thought which serves to suppress and even eliminate that which does not conform to its mechanistic principles. Thereby in the everyday the persistent experience of human beings becomes shackled to the observable and easily recognisable, and everything becomes levelled out, make-able, calculable, explicable and controllable. Whatever cannot yet be mastered is assumed to be master-able, given time and resources, and therefore is already within the instrumental and calculative domain. A deeper truth is thereby unable to encounter the human essence. This impoverishment of the human spirit is the danger from which we are to be 'saved'.

Like any poverty, the rejoinder of spiritual poverty is a hunger. The psyche of the modern human is driven by deep needs, forever wanting something but not knowing why or how or what it is. This spiritual hunger is for something only rarely glimpsed: the replenishment of our vital essence, that we may call 'the soul'. The soul is that unique feature of the human being that is concerned with the 'something more' than its objective existence. Why is it that, despite all the activity of the modern person, the myriad of seemingly endless possibilities, all the restlessness, all the issues that draw our attention, there appears to be a pervading feeling of emptiness, the "what for?" and "where to?" of nihilism. This unfulfilled yearning for deeper meaning in life needs to be attended to; but the question is, how?

Humanity hungers for the uncommon, yet does not know where to look. We seek it in consumer goods and in the fleeting experiences of the exotic. Even the moderately affluent now expect to embark on regular Cook's tours, amassing frequent flyer points and contributing to the injection of 750-milion tonnes of CO^2 into the atmosphere as 250 million tonnes of a non-renewable resource are consumed by aeroplanes per year. We shuttle about the world in jumbo jets devouring the stratosphere to exciting distant lands inhabited by mysterious 'natives' who, as objects on call for inspection by the holiday industry, perform bizarre rituals for our entertainment. Every weekend we join the throngs in our cars in search of "something interesting to do", or "to get away from it all". Well may we ask, "What is it that we are trying to get away from?" As Michael Leunig asks, "Who knows how to stay at home? Who knows how to make a journey into one's own backyard to contemplate and celebrate..." simple things[23]? We seem to have forgotten a sense of what it means to be an earthling and how to become at home. This homelessness has now reached a tipping point where the thoughtless slide into spiritual and environmental disintegration must be reversed. Hannah Arendt names the crisis of

---

23. M. Leunig, op. cit., 'Is your journey necessary?'

human culture as being the "risk of forgetting" that follows from the absence of a mind that can inherit and question; one that cannot reflect and remember what it means to be. No longer able to "recall" and "ponder forgotten things"[24], through its logic of calculability and emulating its own constructions, human consciousness has become 'technologised'. The modern technocratic obsession with order, correctness and efficiency forgets that "all thinking demands a *stop*-and-think"[25].

There is a need to pause and question the role of the pervasive instrumental view of modern life in driving contemporary problems. Of course modern science and technology have provided much that is of immense benefit to humanity and should continue to play an essential practical role. However, they have also played their part in the escalating and increasingly pressing global problems we now face. Yet, the omnipresent instrumental attitude assumes that acquiring ever more knowledge is the only way for humanity to tackle these problems. This assumption is irrational and unsustainable. Humanity seems to be under a massive, institutionalised delusion of being in control and able to fix anything with the tools of science and technology. The dominant instrumental global epistemological mindset is not underpinned by, what we might call, wisdom, and therefore can only lead to more problems. What it *is* underpinned by is a mindless commitment to endless economic growth. The current global financial crisis painfully demonstrates that such growth is ultimately totally unsustainable. Nietzsche, before Heidegger, already understood that what is really needed is to find the means to change the way we live. A mechanistic worldview cannot do that. A holistic approach calls for a radical transformation in the way we approach the looming global environment catastrophe. It involves a departure from the kind of rationalist thought that fuels science and technology, not in order to diminish such thinking, but rather to enrich it. There is a need (at times) for a stepping back from it. Human and planetary welfare is now massively affected by the power of science and technology to equip our activities. Surely, those actions need to be much better considered than only basing them on ever more information and technological know-how.

Human beings clearly belong to the earth and we can either be at home *in* it, or alienated from it. Authentic earthly habitation cannot occur in an indifferent tandem of human beings with the world as some inert backdrop. It cannot be merely an objective spatial being next to each other. To be 'in-the-world' in a 'homely' way is similar to, for instance, 'being *in* love'; it is a state of being. It *in*habits the world, in an attentive manner that grasps the significance of the relationship. Because this significance has been forgotten our activities within the relationship are no longer sustainable. Therefore there is an urgent need to find a way of being that is able to

---

24. OWL p.165.

25. Arendt, Hannah. 'Thinking', The life of the Mind, New York: Harcourt Inc., 1978, p.78.

act responsibly *in* nature and *as* nature. The necessary preparedness to change the way we live comes from a deeper source; one that re-visions the human relationship with Nature as such. Where are we to look for this starting place, this source that seems to involve our fundamental psyche, sometimes named the human soul?

***

I have suggested that in the modern world the human essence seems to be adrift and in need of saving. To save the human soul from its unhomeliness calls for a homecoming. A home needs a suitable dwelling ground on which we can build properly. To seek this ground is the challenge taken up in our philosophical venture; a venture that is an exploration into an unfamiliar realm where progressively light is shed on what is needed for a homecoming as authentic inhabitants of the earth.

> *"The impoverishment of the natural world leads to the impoverishment of the human soul. To save the natural world means to save what is human in humanity"* (Raisa Gorbachov).

"Saving the natural world" has something to do with saving our very psyche, i.e. with the human soul. As Raisa Gorbachov suggests, the impoverishment of nature impoverishes the human soul. There is a link between the imbalance in human nature and the increasing unbalancing of the natural world. There are questions to be considered about this link, such as whether saving nature can actually save the human soul. To restore a functionally adequate planet does not guarantee such a transformation unless it is underpinned by something more foundational, which at this point I have simply named wisdom, or 'spiritual wealth'.

The philosopher John Armstrong suggests that 'spiritual wealth' encompasses the essence of beauty. Moreover, he argues that the converse of this wealth, spiritual poverty, equates with ugliness[26]. The essence of ugliness and beauty has little to do with subjective opinions of what is ugly and what is appealing. Spiritual poverty pursues the obvious and the fleeting; it attaches itself to the materialistic and self-promotion. It neglects beauty and thoughtfulness, and sensitivity to one's surroundings. It leads to an inability to notice the true ugliness in much human activity. It brings about the denial and diminishment of what is profound in human beings and what is essential in a meaningful human relationship with its environment.

The utilitarian attitude to nature means this relationship itself is impoverished. Therefore, I suggest that the reverse of the above quotation is also true. When the human soul is impoverished, when *what is truly human in humans is forgotten*, the consequent all-encompassing desire to control, subdue and exploit is consistent with a diminished view of the unique human place and role within nature. A pervasive ugliness then inhabits the relationship. We need to address this cycle of impoverishment in order to

---

26. 'The Line of Ugliness', *The Age*, A2, p.14. For a more comprehensive discussion on the many facets of beauty see Armstrong, J., *The Secret Power of Beauty*. London: Allen Lane, 2008.

retrieve a ground of significance for human dwelling. There is an uncanny quandary here: to save nature requires the restoration of the human soul, and to accomplish this, 'nature' must be retrieved. To unravel this association involves a much deeper understanding of what is usually described as 'nature': a human construct that is somehow more 'natural' than ourselves. A transformation in thinking is called for that holds a vision of Nature unburdened by worn-out interpretations; one that encompasses *all* of being. This also entails a radical overhaul of customary interpretations of spirituality, the human soul and the essence of things. These concepts will be revitalised from their withered circumstances as the pathway is ventured in the coming chapters.

Existential issues concerning the foundations of ethics, human-nature relationships and questions about what it means to be human in the world are natural subjects for scrutiny by philosophy. Although contemporary academic philosophy can appear abstract, as a quest of fundamental inquiry about the human essence and role it may also illuminate hidden assumptions and inspire new ways of envisaging the world. Ultimately, philosophical thinking should make a *difference* in the full experience and understanding of human existence. Yet, for Heidegger the kind of thinking that he regards as 'true thinking' has to deny itself conclusively answering practical and ideological questions; it has to be 'modest' and not claim to be able to make pronouncements and guidebook instructions. The powerful position of the sciences and the technological mind-set readily alienates and belittles such thinking. Nevertheless, here we engage this thinking as it holds the promise of an essential wisdom. Moreover, as it supports the transformed understanding of how we are to live, we do not stop there. Such thinking is attentive to the guiding metaphysical, cultural and disciplinary presuppositions that operate, often unnoticed, in contemporary worldviews. So often, the constructs that are supposed to explain everything do so by obscuring all details and shrouding the essence of what makes beings unique. We need to think more deeply about what is usually taken for granted; not only to explain things better, but also to show that we thereby find ourselves on uncertain and perilous terrain. Going beyond the presupposed is a worthy task for human beings. Philosophy thereby remains faithful to its traditional role of being a counterbalance to the overindulgence of the age. It provides frameworks for thinking about the issues on which we have to decide. Here we aim for more than a mere compensation for modern excesses, but engage a radical transformation that disrupts the prescribed ways of thinking for the culture. In this disruption and disorientation buried meanings may be uncovered which may well re-enkindle a renewed sense of gratitude for the gifts of the earth, an awareness of its dynamic manifestations and the integrity of the unique moment. This then becomes a journey that replenishes and nourishes the desiccated and impoverished soul.

Modern eco-philosophy has been somewhat pre-occupied with psychology, which has a predisposition to address the subjective position of an objectified person. While psychoanalysis can be a valuable tool, it tends to reduce humans to structures of urges and impulses, instincts and empathy. Heidegger held that, if the *being*, or the soul, of the human being is disregarded, such analysis and therapy "at best, could only result in a more polished *object*"[27]. This book attempts to overcome the subjective and objective position. This does not mean it disregards or tries to erase it; personal dispositions are a reflection of who we are as individuals. Yet here we want to address the foundational essence of human beings, via a transformed thinking with a lessened emphasis on how we 'feel' about things, as regards the relationship with 'Nature' as such.

The last few centuries in western philosophical thinking has witnessed an explosion of diverging strands of interpreting 'reality'. The different schools of thought such as idealism, realism, relativism and constructivism, modernism/anti- or postmodernism and humanism result in complex and tangled interactions between these and in much heated arguing between their various proponents. I strongly believe that none of these are necessarily wrong or right, but rather that all these in some way *connect,* as there is something deeply fundamental that is common to all. It does not mean that we can have it all ways at once, in that every viewpoint is right. However, what is needed here is not to 'prove' that one is right, the other has 'serious errors' etc., but rather, *how* do we connect and disclose the underlying truth these hold, in a way that opens up new insights. These are to *hint* a way forwards on *how we are to live* in a world where the ecological predicament forces us to make decisions.

Today, given the current plight of the planet and perhaps partly because of the seemingly incomprehensible nature of much academic discourse, it seems that popular opinion regards philosophers as amongst the most expendable of all academic species. The spectacular technical, practical and knowledge gathering success of the sciences makes thinking in the philosophical sense appear ever more redundant. This has also become the dominant position within the echelons of academia. In the rush for technological innovation, marketing opportunities, 'integrated' industry partnerships and 'prestigious' projects, who needs crusty old curmudgeons secreted away in dusty little offices pondering airy-fairy notions about the meaning of life? Yet, if the days for the inhabitation of such specialised niches are numbered, we may well ask where that leaves us. I suggest that it may leave us with even larger numbers of 'highly effective, groundbreaking, next-generation' scientists and researchers in large shiny well-organised habitats, who nevertheless find it difficult to see beyond their respective boundaries.

Let me take an example. The university that spawned my PhD recently, in another celebrated "collaborative partnership", began research with their Chinese counterparts in trial cropping of plants that take up soil pollutants and those with soil-binding and soil-improvement properties. Now, I am not for a moment suggesting that this not a most worthy project able

---

27.ZS p.215.

to contribute to the understanding and amelioration of ecological problems. However, at the end of the day, don't these come down to some very well-known primary causes? Why are 12 million tonnes of grain per year and 10 million hectares of Chinese arable land affected by heavy metals and other pollutants? Why has the Mongolian (and Australian, etc.) landscape been disastrously cleared of vegetation? Although the details are complex, the fundamentals come down to: greed, lack of political will, power, poverty, inequality, self-importance, short-term technology 'fixes' and the commitment to growth, to name a few. We can tinker on the edges and at the tail-end of these problems, but that does not solve them at their origin.

To solve these primary causes is not only a matter of ever more information or of using technology in smarter and more innovative ways. In fact, the technical knowledge to understand and solve these primary causes is for the most part already available. However, they also require us to consider matters such as *how* we are to live on this planet in a genuinely sustainable way; one that considers the health, the beauty and the complexity of the earth and all its creatures and species, not only the survival of the human variety. They call for us to better understand what 'makes society tick' – maybe a little social de-construction would not go astray. They might make us question what makes for an ethical approach to land use, food production and distribution. Perhaps we may even reflect on the foundation of all these: on what it means to be a human being as an authentic dweller of the earth. Are we simply ensconced on the top of the species' pyramid with everything around and beneath us for the taking, to be extracted, consumed, manipulated, stored, traded and degraded at will? Is the absolute mastery of submissive nature the only way forwards? Do we try to bend nature even further to try to support problematic lifestyles, and thereby are likely to generate new environmental problems? Or is there another way to envisage our relationship with the earth? Could there be 'something' deeply meaningful about the human relationship with the natural order, something that if grasped might actually transform the ways in which we correspond with this earth?

The 'something' that still seems to be missing, is a basis, a ground from which to build a love for nature; not as something separate from ourselves, but as something into which human beings are seamlessly interwoven. This is the 'uncommon' for which humanity hungers in its impoverishment. Our existential and metaphysical foundations are matters that cannot be dealt with solely by instrumental means. To move nearer to this ground requires a kind of 'stepping back' from the restless corporate and everyday world and recovering questions of human worth and significance as embedded in nature. It might even retrieve a role for philosophy in its perfectly 'rational' and proper task of loving wisdom to help humanity learn how to live. Now, this is not just a role that is assigned exclusively to those few that still get paid to do such thinking! Every thoughtful person, for whom a wholly materialist, reductionist, mechanistic and rationalist worldview is inadequate to properly express human life, can embrace such a task.

"It is one thing just to use the earth, but something quite different to receive the bless-
ing of the earth and to become at home in the gift of this receiving in order to guide and
shelter the mystery of Being and watch over the integrity of the possible."[28]

Like Nietzsche before him, Heidegger here hints at the need for a transformation of the human-nature relationship. He seeks a different truth
from everyday rationality; a transition in thinking that runs wholly counter
to the common sense of the day. Human beings need to learn not to overstep, but dwell in, their possibilities. This preserves the quiet law of the
earth in its self-sufficiency and saves the very essence of what is possible
and what is not in human thought, action and experience. Nature may 'cry
out' at us, but it does so from its fundamental nature, and therefore from
its own terms and concerns within an order of balance and reciprocity that
signal the unitary relationship between the ground of human dwelling and
the earth; between human beings and Being.

---

28. EP p.109; slightly modified.

Part 2

Calling

Chapter II
The Threshold – a Pathway for Thought

The hunger for an uncommon sense of being at home signals a kind of 'call' that originates from the ground of human dwelling. We have seen that everyday life has this uncanny tendency to be overcome by an instrumental view of things which robs it of significance. The call is for the retrieval of a sense of awe and wonder at existence itself, whereby the extraordinary may be perceived in the familiar and usual. Such a transformation of thinking attends to the impoverished soul and reclaims a ground which underpins the wisdom necessary in the face of the contemporary melancholy. To hear this call tackles the unfulfilled hunger for deeper meaning in modern life.

Therefore, it is time to set out and wander a threshold to another way of being. This is a threshold in our thinking about things that guards us against the danger of becoming utterly fashioned by computation, evaluation and 'practical' outcomes. It is a threshold which ensures that another, more thoughtful, poetic and truly sustainable engagement with the earth remains a lingering possibility.

A threshold is a place of change, a stepping stone that represents the possibility of a shift from one state of being to another. A threshold may simply be a space between a building and the exterior; modest and overlooked, simply signifying a change in our physical situation. Or, more strikingly, it may invite a transition in our being as a crossing from a familiar space to a realm that is unknown and yet to be revealed. It is then a passage of transformation that holds the prospect for something of significance to come to pass. It may kindle a sense of otherness or sacredness. But what is this sense of otherness? Is it no more than a simple emotion; a mental response to the novelty of the situation? Or may we somehow differentiate an event of consequence, a moment involving the numinous that implies 'something more'? Is it a communication with the divine? But what do we mean by numinous or 'the divine'. Where and how does it occur? Is it simply a process of the mind or is something deeper involved? This is not to be a discussion on whether human consciousness is identical to physical brain states. Human freedom and creativity are real in some sense, no matter what we believe about their precise nature; otherwise it would not be possible to study them objectively. Of course, a sense of the numinous takes place 'through our bodies', so to speak, yet I suggest that there is more at stake here than a simple matter of perception. Perhaps it comes to pass in what we may call a 'sacred place'; one that seems 'pure' and elemental. But how are we to explore this seemingly unattainable terrain? What is such a domain; can such an unnameable no-thing be examined at all, or is it destined to be a representation of blind faith only, bringing questioning to a rather enfeebled end? To be sure, foundational sources of meaning beyond or outside the mental process cannot simply be reduced to facts, values or metaphysical representations. Their examination calls for a different approach.

To think the numinous or 'the Other' is sometimes criticised as thinking about an empty abstraction, void of content and anti-reason. The thinking I am trying to approach here is certainly not anti-rationalist; indeed rational thought remains essential to achieve a just and flourishing society. But I do aim to show that another way of thinking about truth is possible; one that must necessarily participate in the play of life if the human psyche is to be attended to in a holistic manner. To prohibit the indiscernible and unmeasurable will not obliterate it; its necessity will endure as long as human beings practice their role as beings capable of reflection. As David Wood points out, to take 'the Other' seriously simply indicates a willingness to think the *limits* of logic and truth as correctness, without presupposing the outcome of such a venture[29]. Perhaps then we may find an opening to what transcends us.

The threshold we are about to venture holds the makings of such a realm of otherness, where its dimensionality presents a means of scrutiny. The metaphor of the threshold[30] suggests a 'beyond', an ultimate ground of human dwelling, or the 'ground of Being'. As a feature of the threshold this ground cannot be analysed like an object reveals itself to science, but it can be thought about, brought 'nearer' and experienced, in an experience that

---

29.TAH, p.119.

is 'overflows' the limit of the familiar everyday. To be sure, such reflections appear to be less clear-cut; the very essence of the insights we are seeking means we cannot define them as in a dictionary or a manual. Predictably, we thereby run into terms that mean different things to different people. Yet, expressions such as the sublime, the numinous, the holy, sacred, the divine and the soul all imply a realm of spirituality or of transcendence, an experience beyond the boundary of the familiar. They speak of an otherness that unfortunately too often has been usurped and monopolised by tradition and religion. Jealously held by its custodians, and largely unexamined and unchallenged, the mystery of the otherness thereby dissipates; its residue hardening into the doctrinal edicts of its possessors. In the contemporary world, we too often only hear the voices of those for whom absolute faith and control dominates their worldview, or for whom scientific rationalism has the last and only word. Of course, there are also others who feel the worlds of religion and science can happily co-exist and indeed complement each other. As these apparently opposing realms may, in their essence, not be all that distant from each other, this is not really surprising. Although the latter, unlike the former, is subject to considerable scrutiny and therefore expresses itself with superior rigour, they both honour representation and objectification above all, as the exclusive expressions of meaning.

As Richard Rorty insists, *empirical* truth is not 'out there' disconnected from our beliefs and language. No phenomena as such can be known as these are always interpreted and constructed 'subjectively'. Yet, science claims to be independent of subjective applications by way of its *method*. In the modern era science has used the special, highly effective powers of this method to study a kind of reality with astonishing results. The truth of science is based on the correspondence between propositions and phenomena, and is grounded in the correctness of affirmations. Modern science, in its insistence on verification and its preoccupation with methodology, is solidly based on the interpretation of truth as *certainty*[31]. It attempts to present a solution to subjectivity by showing how *its* language reflects reality and therefore can reconnect to the world through its purification and formalisation. Knowledge can only be held to be 'true' when obtained empirically through its objectivity and method, and published in carefully scrutinised texts. Yet, the sociology of scientific knowledge by Thomas Kuhn has demonstrated that there is no such privileged objective realm. The conditions thrust upon this knowledge constrain it into familiar patterns, folding it into preformed packages that are far removed from reflection and poetic imagination. Clearly the empirical claims to the truth of

---

30. Michel Haar introduces the thought of the threshold in his exploration of the poetry of Rilke and Hölderlin, which he carries out in the light of Heidegger's thinking (SE, p.152-4). Heidegger also writes about the "threshold" of Trakl's poem (PLT p.194ff).

31. Foltz, Bruce V. *Inhabiting the Earth: Heidegger, Environmental Ethics, and the Metaphysics of Nature*, New Jersey: Humanities Press International, 1995, p.76.

science, that is truth by way of logic, cause and effect, located in positively existing, measurable objects, is one particular and important kind of truth. However, it does not tell the whole story of 'truth'. Indeed, foundational truth, as the 'mystery of Being', is resistant to number and measure, to image and form.

Following the 'death of God', neither faith nor an exclusively instrumental view can be an adequate expression of the richness and potential of human life-experience. Therefore here I take a different course; one where an unfamiliar manner of philosophical questioning provides the possibility for going forwards. It attempts to overcome the metaphysics of representation, where God is merely the 'highest being'. Like Heidegger, I try to achieve this by a shift, or a turn, from being solely guided by objective beings to a grounding by the truth of Being itself, as that *source* which is indefinable, yet most elemental. Discerning such a foundation becomes a venture that traverses a threshold, as an overflowing taking place at the *limits* of everyday thinking, the borders of representation, logic, calculation and ordering, thereby enabling the poetic imagination. Of course, this course also holds the risk that new insights may emerge with which we may not be so comfortable, as they may expose the brittleness of some of our most dearly held beliefs. Yet, human development and renewal in thinking is often born from the pain of transition and rekindling, which involve some kind of a struggle and a starting afresh, i.e. a 'new beginning'. To recover the authentic experience of human existence as such, thinking about the concept of the threshold also needs to wrest itself from an inflexibility that restricts its possibilities. To venture the threshold is to indeed approach an 'other beginning'; a realm that seems more open to new insights. We will have a closer look at the nature of these 'beginnings' shortly.

The threshold is an appropriate metaphor to help us explore the realm of otherness, now left vacant by God's departure. There, in a region of passage or transition, the emergence of a different way of being and thinking displaces or interrupts the familiarity of habitual thinking and experience. We have seen that the everyday can become a thoughtless and instrumental discourse that hides a different truth and a particular *kind* of experience that nurtures the wholeness of human beings. This truth concerns itself with the most fundamental sense of existence: "the very soil of our experience"[32]. The region of transition I call the threshold is the unfamiliar domain where language itself speaks from this prolific 'soil', beyond mere words. The threshold identifies and sustains the uniting difference between two domains: between the sensible and the suprasensible; it is where the purely sensible and obvious is transcended. It is an ineffable realm, a region whose topology, its limits and boundaries, seems out of reach and beyond our control. Yet the threshold is 'thinkable'; it provides a pathway for thought to move from the everyday to a thinking that seems 'deeper', more unbounded and concerned with a truth that is more

---

32.TWH p.182.

fundamental that correctness. Once this is understood and experienced, one has founded a 'self', and the everyday may be lived in a more genuinely grounded mode. This then is the role of the threshold: to enable and to enact this experience, which has as its soil the open ground of the existential void beyond.

To introduce the thinking of the threshold we may identify three underlying steps in the participation in this course:

threshold thinking

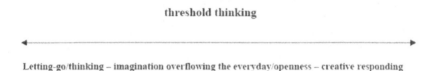

Letting-go/thinking – imagination overflowing the everyday/openness – creative responding

It begins with a need to become aware of the way objective and instrumental everyday thinking takes hold of our very being, thereby diminishing the human potential and responsibility. It becomes attentive to the impoverishment of the human soul. When we are aware of this prerequisite and let go of the desire to know and control everything, the imagination is able to exceed the limits of the everyday and begins to sense the otherness of the uncanny ground of Being. This pathway is not exclusively linear; its experiences overlap and interact and open up the possibility of genuinely creative responses.

Of course, this prescriptive 'recipe' is still insufficient and a rather vague basis from which to build a radical and enduring re-visioning of the human role and place in nature as a whole. It may sound as though we are onto something here, but as we attempt to think our way forwards we are left somewhat adrift in a settling fog of intuitive feelings. Yet, by trawling through Heidegger's difficult works we are able to find some real pointers. As I declared earlier, he tells us where to look and we need to step into the fog if we are to truly live! We shall see that in the realm that is approached in the threshold we cannot *prove* anything and we cannot explain everything to the end, but we can *"point out a great deal"*[33]. Therefore, to sojourn this unfamiliar crossing, the time has come to explore the logic of another beginning in thinking.

Another Kind of Logic

As suggested, embarking on this venture requires a different stance from pure rational logic or correctness, but not one that is irrational or naïve. Heidegger believed that traditional logic and language, indeed the entire range of possibilities (in the realms of religion, philosophy and the sciences) played out in the history of metaphysics (i.e. what he calls the *"first*

33.ID p.22.

beginning"), could go no further. Hinherently translucent to that which lies beyond its own incompletenesse therefore rather grandly announced the "End of Philosophy", i.e. philosophical systems as we have understood them. For him, the questioning about 'Being as such' itself remained essential, but to enable its furtherance a different approach and a 'crossing over' to what he calls the *'other beginning'* is called for. I shall have a closer examination of how these 'beginnings' interplay shortly. The thinking of this other beginning is rooted in what Heidegger saw as the necessity of a "New Philosophical Logic". This is a thinking that retrieves the essence of the first beginning, and what is truly great in it, by questioning it more directly. The "New Philosophical Logic" sustains what I elaborate as 'threshold thinking'. It helps us understand the topology of the threshold, which is deeply implicated with thoughtful and creative imagination and expression. The threshold and its thinking is a course that does have phases and layers; hence it features a kind of 'dimensionality' or topology. Although this structure cannot be fixed in the usual representative manner, it is a *space of significance* that *can* be examined.

To explore this "New Philosophical Logic" and its role in the pathway of the threshold I draw particularly from two sources that shed light onto its features. The first discerns the seed in Heidegger's earlier thinking, a thinking that will grow into the second: the astonishing enactment of his crossing between the beginnings as expressed in *Beiträge zur Philosophie (Vom Ereignis).*[34] The latter contains within it an extraordinary, fugal-like structure in which the 'fugues' perform five unifications that are offered to us as human beings enabling participation in the crossing of the between which I represent as the metaphor of the threshold. I shall examine these gifts that make possible moments of deep significance later, but let us begin with the first source.

This is a little known commentary that provides a most constructive basis to introduce the features of the "New Philosophical Logic" that accordingly throws light on the pathway of the threshold. The first English printing of Heidegger's *The Metaphysical Foundations of Logic* is the only edition in which the translator's (Michael Heim) Introduction appears. His Introduction gives us a clear hint of the changing course and nature of Heidegger's ways of thinking, and moreover what such a model of rationality may look like. There, in a remarkably astute summation of the "New Philosophical Logic", Heim provides a point of entry into the pathway of the threshold, as he alludes to its fleeting phases, layers and boundaries:

> "...*such a logic-in-the-making would emphasise the place or topos where two or more different worlds meet, where an exchange takes place over the gap of mutually divergent domains of meaning and involvement. Such a logic or logos would be self-opening and inherently translucent to that which lies beyond its own incompleteness.*"[35]

---

34. The first English translation is published as *Contributions to Philosophy (from Enowning)*, trans. P. Emad and K. Maly, Bloomington: Indiana University Press, 1999. From here on I refer to it in the text as *Beiträge* and as CP in the footnotes.

These chapters unwrap this statement, as it is precisely this threshold domain, this 'in-between' region of transitional and transformative thinking, which cannot be absolutely defined or located, that is so essential to the becoming at home in earthly dwelling. Let us have a look at the main points of the above statement that draw out the features of threshold thinking.

1. *It engages a logic-in-the-making that is self-opening.*
2. *It is the place or topos where two or more different worlds meet and an exchange takes place over the gap of these mutually divergent domains of meaning and involvement,*
3. *and its thinking is inherently translucent to that which lies beyond its own incompleteness.*

Now, to give us a sense of what we are about to explore, let me expand these just a little. As a pathway of a new dimension of thinking, they imply the following main characteristics to be found in the threshold:

1. In moving from *everyday* rationality there is an unfolding transitional logic that is neither 'illogical' nor traditional assertive logic, which it *underpins*. There is awareness that its thinking will always be incomplete, as its logic takes in a *different conception of truth* from empirical correctness. This logic is not 'willed' by our own doing, but as it self-reveals out of its very *ground* it is communicated from outside the domain of human ordering and representation.

2. It is an *'in-between region'*; a meeting place of different domains of thinking. These different realms reflect dissimilar understanding of meaning and experience; it is a gathering place of the representational thinking of the everyday and that of, e.g., contemplation or poetry, literature, art and music. There is a boundary or limit between the 'dimensions' of thinking, where thinking undergoes a transition. The different spheres 'interact' or 'interplay'; they are not absolute, self-contained or sequential. Time and space interplay.

3. This *ground*, referred to in (1) above, is that inexhaustible source for thought that is unlike that claimed by religion in the form of an ultimate being, as it cannot be possessed, defined or represented. It also always leaves that thought incomplete. We cannot claim that we have reached an ultimate state of knowing. It is not completely crystal-clear; there is no access to some state of unambiguous, pure enlightenment. All we can declare is that beyond its incompleteness is the mystery and truth of Being itself, which is unlimited and 'throws light' on this in-between region, when the everyday barriers to such thinking have been removed. Therefore one can describe thinking here as becoming 'translucent', i.e. semi-open to a different kind of truth; to the uncertain ground of Being and to its unfamiliar manner of revealing.

---

35. Heidegger, M., *The Metaphysical Foundations of Logic*, trans. Michael Heim, Bloomington: Indiana University Press, 1984, p.xii.

The principal trait of (1) concerns the nature of thinking about the '*other truth*' that is encountered in the threshold. This is an encounter where the openness of human beings, or the faculty of conscious awareness that is able to reflect, provides a 'clearing' for the unfolding essence of this truth. This 'self-revealing' of truth is the origin of the incompleteness of threshold thinking; i.e. it can never be brought to conclusion.

The characteristics of (2) emphasise the *topological* aspects of the threshold. They speak of overflowing and surpassing limits and boundaries; sojourning crossings and betweens, and the interplaying of all these. Heidegger placed a great deal of weight on the place of human beings in a 'between' (to which we properly belong) as a kind of clearing where the 'light of Being' can illuminate the primordial truth that is other than, or prior to, objective logic. This 'between' is elucidated in these explorations of the threshold.

The first two attributes allude to the third, that is, to something beyond the threshold: a ground out of which truth reveals itself. The 'death of God' puts us face to face with this ground. Yet, we shall find that this ground is an *abyss*; it 'falls away' as soon as we think we have grasped it. Therefore, threshold thinking always is partial; it is *translucent* to such truth. Now, in everyday thinking this is intolerable: there *must* be an answer! It *has* to be possible for the questioning of something to come to completion, even if it takes an Almighty or another Einstein! Yet, threshold thinking relinquishes such willing. Not only is it prepared to let go of the desire for the *possession* of truth; it thankfully awaits it as a genuine gift.

So, the third characteristic above hints at the uncertain *ground* of the human essence, envisaged as an abyss. It holds both the existential void that so often expresses itself in the fear of finitude and the consequent fleeing into all sorts of momentary distractions. But it also shelters an inexhaustible truth to which human beings inexorably belong; a belonging that itself is deeply meaningful. The ground that is like an abyss is central to a non-theist spirituality, so I shall return to it at various points.

Firstly however, I will have a further look at the kind of rationality that ventures the 'other truth' encountered in the threshold. The first point above states that the logic that takes in this unfamiliar conception of truth is unlike self-assured everyday rationality, but rather is an emerging transitional and transformative logic that is not illogical, but rather 'overflows' practical common sense.

The logic that comes nearer to the other truth - encounters

Here we are concerned with a truth other than correctness of affirmations; one that is not so readily available for scrutiny. It is approached via an 'event' of consequence at a threshold of thinking, where the imagination exceeds the everyday. Heidegger is less concerned with the absolute knowledge of things than with the *basis* for the truth of such knowledge, and moreover with the *union* of truth and Being, which takes place in thought.

It is the *relation* itself, the *belonging* of Being and thought, which is deeply consequential. George J. Seidel refers to this when he writes that it is not "the here and now of present perception and experience, nor is it the accumulation of past knowledge and future expectations, but rather represents man's purest attempt to bring to pass the union between Being and truth"[36]. In such a domain, thinking is in excess of what is immediately present to the senses. Human beings, being capable of this kind of thought, can thereby escape a relativism whereby all truth is ultimately only the subjective present state of knowledge and its future expectations. In thought we can *decide* whether to go beyond the temporal everyday working knowledge of things and approach the truth of being. Yet, this union of truth and Being is not of our own making. Indeed, as Michel de Beistegui suggests, Being thinks *in* human beings rather than the persistent conviction that the human being is the source of thought[37]. Therefore, Heidegger writes "We never come to thoughts. They come to us"[38]. This has its correspondence in the remark by the poet W.H. Auden that one doesn't read a book, a book reads you. This suggests that the creative source communicates *itself* in such imaginative expressions.

Heidegger respected and practiced the norms of classical logic, but he also revealed the impossibility of the ongoing questioning of the 'truth of Being' in this manner. As these norms reach a limit a transformed attitude is called for if the pathway of questioning is to continue. This then becomes a venture, which begins at the *limits* of everyday thinking, i.e. at the limits of representation, logic and calculation and ordering. Yet, in this process we cannot obliterate these limits, and furthermore we need to be cautious not to fall back into a form of naïve thinking. On the other hand to believe that everything is knowable continues the illusion of the equivalence of knowledge and self-consciousness. Knowledge is limited by boundaries that cannot be shown by objective human understanding. Its inadequacy cannot be proven but can only be *experienced*, whereby we come to realise there is always something more that is unthought and discoverable. The idea that the acquisition of knowledge is all there is to thinking is a modern fantasy. It ignores a truth that is other than empirical correctness, but necessary for a holistic human existence; one that offers the foundation for earthly homecoming and genuine dwelling. As this is not the truth discovered by theory or experimental evidence, but one that is revealed and testified to by the totality of human experience, it involves a sort of deep insight, a moment of real significance to which we turn our attention here.

We can think of the unfamiliar logic underpinning the thinking that seeks this truth as a kind of *translation*. Translation is normally a practice of seeking to establish likeness and correspondence between idioms. However, we know that language can also operate and communicate on a

---

36. BNG pp.43-45.

37. TWH p.46.

38. PLT p.6.

different level, one that is less reversible and continuous. Such a 'beyond' of language as everyday discourse seems to be more concerned with deeper meanings and interpretations and with domains of values and essence, which dictionaries cannot provide or even approach. This is the realm for reflection and creativity by thinkers, philosophers, poets, writers and artists. When language is not only a word-tool of communication, it is enabled to speak from a greater depth, that is, it becomes more *'originary'*. This word, that re-emerges occasionally in this book, means *a-rising*, from its Latin root *oriri*, 'to rise'; going back to origins or beginnings. This does not mean biological or pre-human origins, but rather implies an ontological experience *underlying* that of other, more derivative occurrences. Here it refers to arising from a 'centre' near the ground or origin of language, which precedes its articulation.

Heidegger has been criticised for his unorthodox translations as being hardly recognisable from the original. By objective and scientific criteria Heidegger might seem to be a poor translator. This is because for him the translating of texts such as those written by the primordial Greeks must be an interpreting that in some way approaches this originary language. Although he did not lack the competence for such tasks, he insisted that translation per se, as a copying and transposing of the form within a logic of appropriation and predetermined rules, does not necessarily mean that a *transformation* or displacement in thinking has taken place. Merely adopting a translation of a text does not guarantee that anything has yet occurred beyond the simple replacement of one expression for another[39]. As de Beistegui explains, if we remain at this level, everything is already played out and exhausted before we have had a chance to question the deeper significance of what is being translated. We have not yet grasped the unique *event* of translation itself, as an encounter with the other kind of foundational truth, such as we seek in crossing the threshold.

For Heidegger then, such acts of translation are *encounters* with fundamental thought which may be revisited again and again to explore starting places of meaning that cannot simply be reduced to facts, values and entities, or to fixed, unambiguous, reassuring definitions. When language itself speaks beyond words and concrete communication we let *its* thought come to us. Then, in a struggle to think at the limit of words, we may bring to the work of thinkers something from the poetic imagination, now illuminated by the deep truth that already rests, albeit veiled, within their work. This allows us to begin again and progress; to re-experience, amplify and develop what is already significant, but in some measure hidden in their thought. It is in such a manner that Heidegger attempts to encounter the unthought that is concealed within the early Greek words.

Heidegger, in his ongoing philosophical approach stresses the need for such a genuine encounter with the fundamental thought of thinkers in an event that grasps something of real significance. His 'new philosophy' is to

---

39. TWH p.173.

entail more than the customary argumentation. It is more conversational in character; involving dialogue between differing domains not approachable with traditional logic. Such dialogue may uncover an analogical kinship; a belonging together. It is a conversation that gratefully explores the encounters, rather than aiming at winning and losing of argumentation. We can choose to go *counter to* a thinker, or to instead *engage the encounter itself*[40]. The former (often witnessed in philosophical contests) aborts meditative thinking; the latter engages a thinker's encounter and responds philosophically in a way that is more than a polemical argument by confronting the essence of its truth. I appeal for the *acknowledgment of parallelism* between various ways of thinking by scholars about fundamental questions[41]. In fundamental encounters, to see similarities, relationships, and parallels between domains of discourse means that the dialogue can continue. The understanding of such realms can then be deepened without collapsing one into another, ending up with the illusion of a single meaning, or, as Jacques Derrida points out, "forced into the straightjacket of interpretation"[42]. Sometimes, attempting to appropriate a thinker for one's own ends, results in a process of negation, or thwarts the innumerable possibilities in what is yet to be thought to a monologue. It leaves no room for further intervention and interpretation and closes off further reflection. History has shown that the thoughts of even the greatest minds have ultimately been shown to be questionable and open to further interpretation and development. There should be no closure to philosophical thinking, no delusion that we are 'solving' problems; instead philosophy should be open to infinite possibilities. Therefore, to go *to* the encounter of a philosophical insight allows us to gratefully grasp in new unique ways what is truly significant, yet always 'incomplete', at the core of thoughtful discourse.

Often, philosophers are accused of 'neglecting' particular issues. Yet, this deliberately confines one to an un-winnable position. Unless one examines every derivative or related perspective there is always the likelihood of being 'guilty of neglect'. Clearly, Heidegger's seemingly monolithic quest into the meaning and experience of Being was his primary concern. What we do with it is up to us. Rather than becoming bogged down in forced contradiction or inattention, going to the encounter itself (and this need not be in an uncritical manner) enables a creative revisiting, a grasping what is truly distinctive and important about it and then going further unto an unexplored pathway. What is wonderfully characteristic about philosophy is that the possibility always persists to '*start again*', in a uniquely different manner. This is unlike the physical sciences, wherein the researcher, by and large, takes up where a preceding scientist left off. 'Starting again' does not

---

40. David Wood drawing on Heidegger's own reflections about thinking; (see TAH p.154; WCT p.77).

41. As David Wood does in TAH p.122.

42. Derrida quoted from an interview ('Nietzsche and the Machine', 2002) in Reynolds, J. & Roffe J. (eds.). *Understanding Derrida*, London: Continuum, 2004, p.143.

mean that what has gone before it is discarded. It allows one to progress, to re-experience and to amplify and develop what is already significant in thought. In the experience of fundamental encounters an irrepressible space of significance is engaged and ideas are listened to as though for the first time. In this way, philosophical thinking itself is freed from the threat of being swept away by a purely calculative objective knowledge that denies the immeasurable and undefinable; that which is indispensable for the enduring grounding of human beings.

So, the 'translations' taking place in these encounters require a stance of openness to a different kind of truth. Heidegger discerns his characteristic conception of originary truth within the writings of the primordial Greek texts. The word for 'truth', *aletheia*, is probably the most central in his explications of these early Greek texts. It embodies his notion of translation, as a 'carrying-over' of one's thinking in a moment of transformation. He believed that ancient thinkers such as Anaximander, Heraclitus and Parmenides conceived the great insight of such truth as *aletheia*, without attempting to define or analyse it, which would have resulted in the dissipation of its primordial significance[43]. Instead, he construes it more as an *originary experience* of thinking; indeed, it is a celebration of the possibility of thought itself. This for him is the onset of the '*first beginning*'; the first moment in the unfolding history of truth, a moment that perhaps occurred "for the first and last time" in this manner in western philosophy[44]. What Heidegger calls the "*beginnings*" refer to two moments in this history[45]. They define the critical distinction between two kinds of truth. We are familiar with correctness, as the customary concept of truth. This is the correctness of a proposition by the correspondence between exchangeable conditions. It is the knowledge that condenses into dictionaries and encyclopaedia, and that seems to belong to representation, compartmentalisation and classification. However this knowledge does not know *the event*. It does not grasp the moment of deep significance from which this knowledge originates. The event, as we shall see, is drawn from the void, the inexhaustible ground of Being. It is therefore beyond the numeration of categories, calculative thought and cannot be signified into language in a straightforward way.

The *other* truth (*Wahrheit*), has its genesis in the *aletheia* of 'first beginning'. Heidegger regards correctness (*Richtigkeit*) as being derivative of this *originary* and pre-metaphysical essence of truth. The other truth (*aletheia*) needs the openness, inherent in human beings, to its unfolding essence, which is sensed in language beyond words. It is an understanding of truth which, in threshold thinking, is recalled out of its forgotten state, sheltered and responded to. The German words *Wharen* and *Wahrheit* have crucial connotations of safekeeping and preserving, and of *freeing*[46]. This 'freeing'

---

43. P p.ii.

44. TB p.52.

45. TWH p.170.

by the other truth releases us from the shackles of an inadequate worldview that loses sight of hope, disables the imagination and stifles creative responses. When it is realised that truth is not only an issue of concurrence between contention and its object, we are ready to grasp the primordial phenomenon of truth; what may be called 'the essence of the true'. This truth reveals itself in the fundamental freedom of the open realm of the threshold, where human beings are themselves open to receive it as a gift.

As a demonstration of this conception of translation Heidegger insists that to simply translate *aletheia* into its contemporary equivalent does not convey its primordial sense, the truth sensed in the inception of the first beginning. In *Parmenides* (pp.10-14) he writes that merely *replacing* the usual literal translation of the Greek *aletheia* by "truth", or even the cumbersome English translation "unconcealedness" of the German *Unverborgenheit*, is inadequate when applied to the Parmenidean originary experience. He emphasises that although "unconcealedness" approaches the truer meaning of the word, if we simply remain with such a word exchange, it still does not "*carry us* over" [*uns* über*setz*] into the manner and realm of experience of which Parmenides speaks[47]. We are still assuming correspondence as convertibility and therefore for Heidegger "we are not yet actually translating". Merely constructing an opinion out of a proposition does not mean that anything of significance in thinking has necessarily occurred. Following thinkers or poets by going to their own encounters of thinking is not simply a matter of understanding their philosophy or poetry and to recapitulate it. We would then still be 'dealing' with truth as correctness and with words as tools, even when they are unfamiliar such as "unconcealedness". Conventional literal translations cannot *by themselves* displace, deport or uproot us from habitual ways into a more fundamental experience. To encounter the other truth involves a transformation, which the customary linguistic act of translation as '*over* carrying' (*Über*setzung), the exchange of expressions, does not automatically achieve. We are therefore looking for a transformation that is a "*carrying* over" (Über*setzung*) or a "*transporting* of ourselves into a new realm of meaning"[48]. It is an experience that is not so readily brought to language, but may be manifested in transformed ways of thinking, writing, acting, creating and being. Indeed, Heidegger is hoping that in such a deep transformation the thinking of the essence of un-concealedness can once again, even in the modern world, "become the glowing fire of the hearth of authentic human being's existence"[49] .

So here we are concerned with such moments of encounters into the region and manner of the elemental experience of Being as such. When we are transported by such translation, even words such as "unconcealment"

---

46. QCT p.42n.9.

47. See TWH p.173ff, P p.11ff.

48. 'Translators Foreword', P p.xv.

49. BQP p.127.

or "unhiddenness" that previously seemed meaningless or at least overly abstract, are gradually 'opened up'. *A-letheia* as "un-concealment" basically affirms that all things have a veiled character; we never 'know' them completely, and we need to be aware of that. Perhaps then we can see that concealment itself, as it shelters the other kind of truth, is nothing negative. It holds the enduring possibility for human beings to partake in its awesome play of un-concealing.

As words that attempt to express this primordial truth are not easily representable, what these really mean, in their essence, has not been decided. Instead their deeper meaning must be gained by a 'struggle' that in itself is to be genuinely experienced. The critical distinction between truth as correctness and the Greek experience of the essence of truth involves this kind of (un-wilful) 'wresting' from concealment. This exceeds what we might call 'context', because here we move beyond the words to their ontological and historical ground and the essential originary experience from which these first appear. To sense language beyond words and concrete communication, and to think at the limit of words and yet bring its experience to language, is a struggle. This kind of translation wrests itself from the law of literary exactitude and becomes a movement between spheres of language. It is translation-as-transformation. There is still communication, but not as a spatial and sequential continuum. Rather they are "disjunctive and interspersed with gaps and drops. Between them, there are no easy passages, no ready-made transitions and pre-established equivalencies"[50]. This truly describes the 'in-between' of the threshold as a realm of significance that sustains and transforms. It *displaces* human beings into a sphere of unfamiliar and deeply consequential perceptions that are normally hidden, whereby the usual becomes the extraordinary. The 'between' nurtures the venture of the human soul 'becoming at home'.

Now, this transformative shift is not to be limited to the interpretation of texts, but may also be applied to other sites of human experience, where, in a journey of displacement, situations, themes and dialogues may be carried across the space of their own unfolding. The displacements taking place in these kinds of events unfold truth from a horizon or boundary where something new comes into being. To be 'open' to the ground of that which discloses the essence or 'truth-character' of things is a process that begins by not immediately reducing everything to familiar representations. In reflection these events are not thoughtlessly absorbed or consumed, but experienced in a moment of consequence which 'carries us over'. The experience taking place in this kind of translation ruptures the familiar. One is 'dislocated', allowing the uncanny or the numinous to be sensed. Even familiar things may be re-experienced in a completely different light, enabling yet new creative 'translations'. Then things are returned to the sphere of their most fundamental nature where they can truly "stand and shine within presence"[51]. This may be a moment of insight, an unfamiliar

---

50.TWH p.179.

sense of 'otherness', which *begins* something new in the thinker, in a manner which is not so easily explained by the traditional rules of logic. The events of consequence leave in their wake traces which leave us not only 'more experienced' but which also grant us the lure for reengagement; the possibility for other unique encounters. These traces are an acquisition that endures.

These beginnings are moments of transitional and transformative thinking that do not abandon the essence of things, that which cannot be pinned down, as a lost cause. What it *does* abandon is the path of certainty and the idea of the human project as one of mastery, a path to knowledge where error must be purged without the need to begin again. It discards the customary expectation that whatever has a beginning needs an end. The beginnings are not in accord with ends, but rather come to terms with the human situation of always being in the middle of things. This 'between' approaches a fundamental ground that is *always* essential and consequential for human existence. It is a realm, in modern life largely forgotten, that needs to be frequented and reflected upon as it imparts significance and wisdom into this existence. The obsession with calculative thinking, now accepted as the only valid mode of thinking, has resulted in the indifference to the merit of *reflection*. Its retrieval restores a balance in modern thought. Thereby intellectual understanding and actions are grounded in an affirmation of the *value* of this way of thinking that has become remote: a kind of contemplation, which expresses the essence of the human being. This is the unique facility that makes us truly human.

The reflection needed to approach the pathway of the threshold involves a kind of unfamiliar thinking that is more than a mere making conscious of something. It entails a "mindfulness", which interrupts cognitive activity. We do not yet have reflection when we only have consciousness. In reflection we circle around the "puzzling character of the phenomenon" where we slowly come nearer[52]; at times there is a narrowing of the circle as we come closer to its centre, yet without the expectation of reaching '*the* answer'. Sometimes the circling becomes more elliptical, and we seem to move further away from the allure of our thinking as it withdraws. Maurice Merleau-Ponty also has observed that it is necessary to *step outside our normal involvement with the world* in order to make that involvement the subject of philosophical inquiry. Although, in reflection, we 'step back' from that world in which we are involved, we need never lose sight of that world, which is where human activity takes place. As he wrote lyrically,

> "*Reflection does not withdraw from the world towards the unity of consciousness as the world's basis; it steps back to watch the forms of transcendence fly up like sparks from a fire; it slackens the intentional threads which attach us to the world and thus brings them to our notice*"[53].

---

51. TWH p.182.

52. ET p.197.

He continues that this reflection reveals the world's dwelling ground as "strange and paradoxical". Thoughtful reflection on life experience exceeds the reality presented by the certainty of rationalist science. For Heidegger this 'stepping back' from one kind of thinking to another is not just a change in attitude. In 'Science and Reflection' he writes that it is "the calm, self-possessed surrender to that which is worthy of questioning", which for him is the truth of Being[54]. The thoughtful attentive reflection that Heidegger calls *Besinnung* is a kind of *"concealed thinking"*. Jean-Luc Nancy describes this as "thinking that conceals itself from the anticipation and the demands of [objective] knowing, while still remaining [conscious] thinking"[55]. It is neither rational nor irrational but encompasses an altogether different domain; a 'new philosophical logic' that grasps a deeper truth which nevertheless does not grant 'the whole'. It embodies a shift away from the absorption into a pre-defined world of objects by a subject (who is set over against this world) towards the thinking that is 'translucent' to the truth of Being. Having 'stepped back' one is not simply directed or dictated by beings; especially not those who are driven primarily by ego. In a movement of thinking that is integral to the venture of 'becoming at home', we become more receptive to the very *difference* between the domains of the familiar and the unfamiliar. The strangeness of this difference, implicated in the 'strangeness' of the world, becomes something worthy of attention. This way, Heidegger insists, human beings can be saved from "being lost in the bustle of mere events and machinations"[56]. The history of metaphysics has left behind a lot of baggage that has overburdened us with unpleasant and bewildering images of messiahs, omnipotent judges, clockmakers (setting the universe in motion), weird connotations of 'end-times' and 'Chosen Peoples'. Here in the realm that sets free we encounter the release from loopy constructions to the liberated play of belonging and responding to an essential truth.

So, in Heidegger's enactment of this truth as expressed in *Beiträge*, this 'in-between' of the threshold is the realm of a crossing where the thinking of the first primordial (Greek) beginning is deepened, i.e. brought nearer to a deeply hidden truth by the involvement of the 'other beginning'. This attentive meditative stance is different from the more analytical thinking that grew out of the first beginning: the metaphysics of presence and of objectivity, towards which we continue to be drawn as well as called away from. We are still attracted to the idea of representable Gods who we can appeal to. However deep down we are called to be true to our selves and to our hidden divine impulse that calls us to venture out to the between; to be

---

53. Maurice Merleau-Ponty, *Phenomenology of Perception*, New York: Routledge Classics, 2002-5, p.xv.

54. QCT p.180.

55. FT p.37.

56. CP p.40.

cast out from familiar domains. There things appear in a fresh light and we grasp that which gives worth and meaning. Logic does not fade away to be prevailed over by rapturous reverie, but there is now a union between the realm of reason and that of poetic imagination.

The beginnings signify an emerging logic within a gathering place of different spheres, enabling an exchange of truth to take place in the overflowing of limits. This logic is self-opening as it is innately translucent to that which lies beyond its own incompleteness. The truth of Being arises out of the inexhaustibility of the strange, unhomely realm (explored further in Part 3), the 'pure open' that Heidegger calls the *Abgrund*, the abyss. When, in thought, we reach the limit of everyday things, the imagination senses this pure open space beyond and overflows as a kind of surplus that spills over this boundary. As we shall examine more closely shortly, the limits themselves become translucent as the imagination 'touches' them and is transformed. Here in the 'interplay' between the different spheres and dimensions of thinking there is an awareness that thinking requires an openness, or mindfulness, beyond purely calculative, objective postures, towards the uncertain ground of our being and the unfamiliar manner of its revealing. This is the undecided pathway whose exploration lies ahead. There the possibility for familiar, ordinary things to become 'unique' and 'extraordinary' is anticipated. We shall see that such care-full dwelling on the earth is more attentive to beings, places and moments. In this 'space' the things of the world are something more than mere objects, stockpiles to be quarried, packaged, sold, consumed and disposed. Nor are they the diminished components of mechanistic and orderly systems. Instead their very being has been salvaged from oblivion.

So how do we find the means to open ourselves to the radical unfamiliarity of the experience of the different realm of truth? How do we open up our thinking and our language to the fundamental experience that underlies that of all other? To be aware of the possibility of such an unfamiliar manner of thinking about truth has made for a good start. Moreover, the threshold's topology provides a course in the venture of seeking the experience that is fundamental to all experience and innermost to 'becoming at home'. Examining its phases, structures and layers will shed light on this course and will support us to become more aware as to what is happening here.

Chapter III
The Threshold - a Topology of Consequence

Let us remind ourselves of the second feature of The New Philosophical Logic (p.55) that calls attention to the *topological* aspects of the threshold:

> It is an 'in-between region'; a meeting place of different domains of thinking. These different realms reflect dissimilar understanding of meaning and experience; for instance as that of the everyday and that of contemplation or poetry, literature, art and music. There is a boundary or limit between the 'dimensions' of thinking, where thinking undergoes a transition in order to be able to 'overflow'. The different spheres 'interact' or 'interplay'; they are not absolute, self-contained or necessarily sequential.

Here the venture continues by examining the phases and layers, the margins, limits and boundaries of the threshold region. All these in some way interrelate. Moreover, as we shall discover, the 'openness' of this in-between realm transpires out of the 'pure Open' ground beyond. The image of the threshold facilitates the 'spilling over' of the imagination and helps us go forwards on the journey of 'becoming at home' as earthly dwellers. As a metaphor it conceptualises the nature of the *fluidity* and sense of movement of the 'in-between' of the 'passage'. To venture such a region of thinking, one cannot expect a roadmap, even less the technological reassurance and egotistical affirmation of digital navigation systems, for guidance. Indeed, the demand for unambiguous charts, simple, instant and guaranteed answers to the confusion of the time is symptomatic of the certainty for which the contemporary age yearns in its homelessness. There are no definitions for these un-representable 'regions' of thinking. Yet they are wonderfully 'thinkable' when thought is ready to be engaged in an ongoing operation of dis-location and rupture from the familiar so it may be freed to truly think. This invites forgotten foundational understandings back into the 'site' of their essence.

Firstly, expanding the 'threshold thinking' diagram on p.51, an overview of the threshold may be visualised as three distinct realms that constitute its overall pathway.

◄──►

The everyday - the between (the threshold) - the Ground of Being
The familiar/habitual - Overflowing - The Unthought/the Unknowable

◄──►

In the attempt to envisage it in this way does not mean this is exactly how it *is*. In any case, we are not examining an '*it*' here. The struggle to bring this domain to language perhaps expresses itself in the apparent need for the copious use of 'scare quotes' in its examination. This threshold is not a definable object. Yet, being the kind of beings we are, we need some kind of visualisation to help us understand what is going on. Therefore we do not shun the metaphors and symbols, images and allegories that support us in the venture of the crossing. Indeed, it is in the spheres of creative thought and discourse that take up such means where the distinction between different kinds of truth may be *experienced and practiced*, without necessarily desiring to define and posses. Any attempt at mastering would abolish this topology of consequence. Here I simply stress that the threshold metaphor does not pretend to objectively represent the space of access towards the pure open site of Being. Instead it can only hesitatingly superimpose an image upon it.

The threshold is not a prepared path which we can traverse safely and comfortably towards the recovery of an authentic self. As this in-between is not readily pinned down in space and time there is not a *bridge* here of the kind that facilitates a ready-made transition.[57] Because the spatiality of the in-between of this crossing is unstructured and discontinuous, a *leap* is needed. This is a moment of decision in response to an uncanny 'call' which expresses itself in the incompleteness of the everyday, the hunger of the human psyche and the need for another truth. The threshold is the 'landing' of this leap 'over' the boundary of the everyday into the crossing to this truth. Here our role is not to build a bridge between Being, the numinous ground of existence, and beings, as though they are on opposite banks, i.e. as if they do not already belong to each other. Rather, we are to simultaneously transform our awareness of them, by bringing out the unity and truth of *both* domains.[58] Then beings are grounded in Being. So, there is a *kind* of bridging taking place. As I shall clarify shortly, the 'beginnings', the moments that mark the decisive distinction between the two kinds of

---

57. Cf. CP p.142.

58. CP p.11.

truth, 'interplay'. Via the thinking of the threshold what truly *belongs* together is brought together.

We may already sense some of the attributes of the coming together in the crossing; such as a leap in response to a call, an interplaying of beginning ways of thinking. All these play a part in grounding human beings as earthly dwellers. Heidegger did not simply understand these as the incidental aspects of a more poetic way of thinking; he found them worthy of much deeper reflection. He saw them as gifts from Being itself that unify the event of significance that takes place in the crossing. This coming together of complementary offerings gives rise to the unique event that Heidegger calls *Ereignis*, a moment of existentially deep significance. This is the glowing treasure, illuminated by the 'light of Being' itself that he seeks and enacts in the second source from which I draw in the exploration the threshold: his *Beiträge zur Philosophie* (*Vom Ereignis*). The approach in *Beiträge* may well be brought into play as a source and guide for the venturesome movement of transitional and transformative thinking necessary for 'becoming at home' in earthly dwelling. So, to guide us in this venture of '*becoming* at home' as an authentic way of being we 'leap' into this most unorthodox exposition, which has at times been criticised by those who do not understand its radical approach as obscure, opaque and incomprehensible and without the proper arrangement for a scholarly work. It has indeed a very extraordinary structure, which, moreover, has its parallels in the topology of the threshold.

A Musical Offering: the fugues that herald another beginning

The text of *Beiträge* is joined together according to the musical model of the fugue, whereby Heidegger purposely develops a complex but unified array of contentions. Unlike musical themes, words cannot be counter-pointed, and therefore this written form of the fugue can only be adopted conceptually rather than literally. It has six fugal movements that are united in an underlying theme, but each is developed from a different unifying attribute. He insists that although there is an organisational structure, it is not a "system" that can simply be repeated. Like a musical fugue its structure is to be *experienced* rather than dismembered. Its primary unifying theme is the occurrence of the event he calls *das Ereignis*. For Heidegger this is his attempt to achieve what has always eluded philosophy: "to *grasp* the truth of Being in the fully unfolded richness of its grounded manner"[59]. He does not try to do this in the usual metaphysical way, by way of a representation which stands in the way of such understanding. First he wants to retrieve the experience of the early Greeks in the inception of the 'first

---

59. Iain Thomson was among the first to properly grasp Heidegger's *Beiträge*: "The Philosophical Fugue: Understanding the Structure and Goal of Heidegger's *Beiträge*", *Journal of the British Society for Phenomenology*, 34(1), Jan.2003, pp.57-73, n.12.

beginning'. In a primordial experience that became the source of meta-physics, they regarded Nature in a way that does not seem possible today. They truly experienced the self-emergence of Nature in a manner of awe and wonder that today needs retrieval via the poetic imagination if the human earthly dwelling place is to be grasped. In this particular kind of beholding, emerging nature is perceived *as though for the first time*. Of course, because of the amazement the early Greeks felt before the overwhelming, inexplicable primordial phenomena of the earth and the heavens, the experience of the mysterious opening of existence itself, came more easily to them than for modern humans. Yet, science and technology only *appears* to have made their astonishment redundant. Attesting to the role of this experience as integral to fundamental human nature is the enduring probing of the numinous by thinkers over the aeons.

Heidegger not only attempts to experience the primordial Greek event, but wants to probe its meaning and implications from a new stance. This shift or "turning" in Heidegger's pathway of questioning was worked out from 1936-1938, culminating in the perspective published in *Beiträge* and other works around this time. Here thought crosses into the realm where human beings are attuned by the truth of Being itself. In *Beiträge* he attempts to overcome the limits that metaphysics has imposed on itself. Metaphysics seeks the truth of Being, however if Being is asked about in such a way that it can only appear as 'something', such as the highest being, usually called God, the unity of truth and Being is extinguished. In *Beiträge* he tries to overcome the language of metaphysics, not in a way that surpasses beings, but by turning this transcendence back to the grounding *source* of human beings, from where self-disclosure, or the emerging of nature, takes place. The notion of transcendence becomes that which summons us *to* the ground and our belonging to it. The source is grasped, not as a ground as such, but more in an active sense of *grounding* that gathers the, rarely encountered, unity residing in this in-between realm. Following his "turning" Heidegger now emphasises the determination of transcendence in a more originary way where one 'steps beyond' the ego, allowing the authentic human essence to unfold, providing a clearing where the truth of Being may be glimpsed.

Beiträge marks a radical re-visioning of Heidegger's probing into the meaning of Being and lays the *foundation stone* to all his later work. We can describe it as *transitionary*, not because *Beiträge* is incomplete, not fully developed, a work under construction or inconsistent, but because it is an attempt for the *undergoing* of a transition, where there is a *'carrying* over'; a translation in the sense I discussed earlier. This is a transit out of metaphysical ways of thinking to a more *originary* way of thinking about Being as such. This movement in thought is expressed as a transitional crossing: like our threshold, it has the "character of a passage"[60]. It is a passage that starts from metaphysics, exceeds it and yet points back to metaphysics. In

---

60. CP p.328.

a sense, *Beiträge* itself is also a *preparation*; as the way it grounds human be-ings is an essential feature of a transition which cannot be concluded. That is why its thinking is always a crossing, an in-between or a threshold. Its manner of thinking does not *expect* completion. It is both transformative and transformed. It is a thinking that has moved away from its familiar ob-jective and utilitarian mode and continues to transform, i.e. make us 'more experienced' at living as authentic human beings. Here, in moving beyond success and failure, it cannot be claimed that in *Beiträge* Heidegger has suc-ceeded in his quest to grasp the truth of Being. But he does uncover an overgrown forgotten pathway that brings us nearer.

An Unfamiliar Event

Ereignis is the event of consequence where thinking becomes transformed by the truth of Being itself. This event cannot be examined like an object presents itself to the scrutiny of science. Although there has been much academic discussion about Heidegger's *Ereignis*, ultimately it can only be *enacted*, and therefore its experience will be different for each wanderer of the pathway. Although the fundamental underlying event called *Ereignis* holds the unity of an originary moment of deep significance, its experience and articulation is unique for every individual. In the 'Preview' of *Beiträge*, he clearly insists that those whole follow him will not duplicate Heidegger's encounter with *Ereignis* or his attempt at expressing this experience. Therefore, we need to become a little more adventurous in interpreting and applying his insights than is often permitted by Heideggerian scholars. He also insisted that those who follow will not actually be able to '*explain*' his attempt, according to the usual rules of logic. We can only approach and enact it as a hesitant *encounter* with 'another', more primordial truth.

We have seen that for Heidegger, to act out the event as *Ereignis* aims to overcome metaphysical representations. No longer will the God of the Holy Scriptures, the picture stories of heaven and hell, of angels and demons, be sufficient. These are the predictable outcome of the human de-sire to conceive the inconceivable. In an (perhaps not entirely successful) attempt to articulate this radical approach, he adopts the archaic *Seyn* in-stead of *Sein* to emphasise that Being is 'no-thing', not a substance or a concept, but belongs to a dimension that precedes all conceptualisation and all rational knowledge. Moreover, because of the transformation in thinking enacted there, Heidegger abandons familiar ways of expression and ventures new possibilities. To express the inexpressible still needs to be brought to language in some way. The unfamiliar language of *Beiträge* is a testimony to his struggle.

While the event that Heidegger calls *Ereignis* has been the subject of much controversy, mystery and speculation, I see it as both simple and complex. It is simple in that it is an experience that does not objectify, ana-lyse or dismantle things into oblivion. Its unfamiliarity rather indicates that this *kind* of deep reflection has become difficult for us. Nevertheless, as David Wood points out, although the kind of thinking it exemplifies is in a way simple, the specific performative thinking *from Ereignis*, in the way that Heidegger performs it, is a highly complex *philosophical* achievement[61]. Although we need not take on Heidegger's almost obsessive preoccupation with avoiding the dangers of misinterpretation, bearing his careful approach in mind can be most valuable. The experience of *Ereignis* should be anticipated and prepared for. Therefore, before poetic imagination and creative responses can occur, a certain preparation or stepping back is called for. Although thinking here cannot be judged by the usual formal rules, it still requires appropriateness of 'style' or rigour if arbitrariness, naivety or worse, irrationality, is to be avoided. To be sure, its philosophical practice is demanding: as Heidegger writes in *Parmenides* (p.121), "Of course, a certain knowledge belongs to this thoughtful thinking as does a carefulness in reflection and the use of words". He sternly warns of using 'throw-away-lines' such as "immortal works" and "eternal values", which avoids having to be precise in essential matters[62]. *Beiträge* and its surrounding works at that time, *Basic Questions of Philosophy* and *Mindfulness* are a testimony to the degree of attentiveness and thoroughness that he applies to this project. *Basic Questions* and *Mindfulness* develop the *articulation* of the thinking that is always present in the background and brought to language in *Beiträge*. However, in our explorations of the threshold, we do not need to concern ourselves too much with the intricacies of the interrelationship between these works and their diverging methodologies. Again, in *Beiträge* he emphasises the need for those following his unfamiliar way of questioning, to lay out their *own* pathway, which for *them* enables thinking about that which comes "from far away", i.e. from what is normally hidden, as his or her own unique undergoing of *Ereignis*. It is up to those who trail him to make sense of it in the experience of life as we find it.

What is particularly remarkable and relevant about *Beiträge* is the *performative manner* by which it opens up new possibilities for thinking. So, what do I mean by performative thinking? It is a kind of thoughtful expression that occurs when there is a particular dimension (only occasionally engaged) in one's thinking practice, where one is 'displaced' by the very realm itself, with which one is concerned. This is a particular kind of 'event' where the one who grasps a correspondence with a realm is *transformed according to the realm* rather than engaging the subjective self. I return to the matter of subjectivity and its importance in relation to the human 'soul' again later, but here I stress that this "thinking comes from the matter that is to be thought"[63]. The ego and the assertiveness of opinions then become less important than the significance of being in the realm itself. Here, in the in-between of the threshold, this realm has its origin in the truth of Being. The realm itself '*enables*' the displacement from calculative and

---

61. TAH p.161.

62. PLT p.77.

63. *Der Spiegel* Interview.

subjective thinking to a less certain, yet deeply meaningful way of being. This is the 'event' that Heidegger calls *Ereignis*. Therefore his chosen subtitle for *Beiträge* is *"vom Ereignis'*, *from* the event that appropriates and grounds human beings.

So how does *Beiträge's* fugal form serve Heidegger's goal of a crossing to another way of thinking? He employs it as a provisional ground plan which prepares for this crossing and for the engagement of the 'other beginning'. It provides a groundwork that transforms and initiates thinking for that beginning. It is not a plan that incorporates timelines and arranges objects and observations in order of priorities or importance, but attempts to break into the free play of time and space. Although it is formally organised into seven sections, we can identify the six fugues (*Fügungen*) as 'unifications' rather than divisions. Heidegger himself insists that each of the fugues stands for itself, yet only in order to express the necessity for their "essential unity"[64]. They describe differing, but interconnected and essential characteristics of *Ereignis*. He expresses his vision of a reverential development of a unified whole by the addition of the seventh section, which attempts to grasp all the fugues at once. The six fugal movements each begin in a similar way as short counter-pointed subjects; these then gradually develop the underlying theme, while at the same time remaining *under the influence of the event that unites them all* (*das Ereignis*). Again, here the performative manner of thinking wrests us free from the hold of wanting to categorise and dominate. Each fugue from which he approaches this event then becomes an enablement, a gift of understanding whereby one can speak *from the site of the event itself*, rather than from the ego. Thinking is then enabled and sustained in a manner whereby "thought unfolds as it is thought". The realm shapes and guides our thinking whereby things are perceived in a way where they self-disclose from their essence, rather than simply observed and evaluated objectively. Things then '*essence*', where essence is understood as a verb, as the active way in which things 'remain in play'[65]. Such 'essencing' has elements of 'living', 'dwelling' and 'being', and of revealing and concealing, as manifestations of Being. Essence (*Wesen*)[66] for Heidegger does not mean some metaphysical reality, or substance, that stands behind things. Rather the essence of something is the centre of meaning, *out of which* we must think its relationships with all else. So things *belong* to essence, rather than *possess* it.

To better understand what happens in the threshold it is enlightening to explore the fugues and their role within the "passage" that joins the beginnings. The six fugues, as unifying "joinings" (the word used in the English translation), are the offerings that enable the transition to the other beginning that thinks Being's truth. They make the in-between possible as

---

64.CP p.57.

65.QCT pp.3n.1, 30-1

66.Most English translations of Heidegger's works have commentaries on Heidegger's use of *Wesen* and its various derivatives.

the transformative transition in thinking, which grounds human home-coming and dwelling. Although the unifications are from different sites of access or standpoints, they all involve the distinctive unfolding of the truth of Being, which ensures their unity. They are as follows:

"The Echo" or recollection (*Der Anklang*)
"The Interplay" or "Playing-Forth" (*Das Zuspiel*)
"The Leap" (*Der Sprung*)
"The Grounding" (*Die Gründung*)
"The Ones to Come" (*Die Zu-Künftigen*), and
"The Last or Ultimate God" (*Der Letzte Gott*).

These provide keys to the point of entry into the transformational and transitional thinking, or the "New Philosophical Logic" of the between of the crossing of the threshold. This realm, between the divergent domains of meaning that we are concerned with, is where the unifying event of deep significance (*Ereignis*) may come to pass. In his 'Preview' in *Beiträge*, Heidegger expresses his task of thinking towards "the riches" to be found in the "immeasurable" and "incalculable" experience of *Ereignis*. As noted, in the move away from calculative and purely rational thinking, this becomes an event where Being truly essences; it enters into the play of an authentic life.

The difficult language employed by Heidegger is *his* unique response to his transformed understanding of this happening. In a tortuous sentence in the Preview[67] he attempts to describe the essential unity of the six unifications as a thoughtful questioning that takes place within and out of the unifications themselves. Let me tease out this sentence as follows:

The first and the other beginning *interplay*; they are not isolated from each other but interact in a playful and uncertain manner. This interplay arises out of an *echo*; a gentle call that reminds us that something is awry in the way we no longer reflect on what it means to be.

Although this unease is usually unrecognised, this now causes a kind of uncanny distress that gives rise to a spontaneous decision to *leap* into the crossing, the threshold of the truth of Being, whereby this truth is *grounded*.

This then is a preparation for those who are to be true dwellers of the earth ('*the ones to come*') and for the possibility for '*gods*' to come to pass.

The above are the unifying characteristics of the event of deep significance that is explored as we move along, but perhaps we already intuitively begin to sense the elements that unify and bring about the *Ereignis*. There are features in Heidegger's fugal argument which are common to all the unifications, such as the lack of noisy, self-important opinions and the movement away from representation. In all of them reverberates a necessary 'openness'. This is a comportment of human beings that is not a psychological disposition, or simply an attitude, but rather a state of *preparedness* for *Ereignis*, the event where Being 'happens'. It is the enduring human

---

67.CP p.6: This thoughtful inquiry occurs "in 'The Interplay' between the first and the other beginning – according to 'echo' of *Seyn* in the distress of *Seyn's* abandonment – for the 'Leap' into *Seyn* – in order to 'Ground' its truth – as a preparation for 'The Ones to Come' - and for 'The Last God'".

'ground of care' that sustains our belonging to Being[68]. It resonates with elements of endurance, reticence, holding oneself back. It means that the 'Leap' away from calculative and everyday reasoning into Being's uncertain proximity (the in-between) is always possible, even within the clamour of the modern mindset. Out of this grounding posture Heidegger discerns specific *guiding* moments that may come into view within the fugues, such as "awe", a flash of insight (an *Augenblick*), a hesitation and "shyness" that may come into view within the fugues[69]. Moreover, each unification highlights its own resonance with Being's elemental influence and experience ("sway" as *Wesen*), which we shall deal with as we continue.

To set the scene of the realm we are exploring it may be helpful to refer to the diagram below. It represents a summary of the six unifications that bring together the Event that Heidegger calls *Ereignis*. They all may be perceived as gifts, *offerings* of the experience of Being, that interplay and enable the spilling over of everyday perception and imagination and the crossing of the pathway between beginnings. Although we know that this structure cannot be fixed in the usual representative manner, there is a kind of unified '*dimensionality*', a space of deep significance, which may be made clearer by the image of the threshold. It involves an interplay of time and space, which characterises the various 'moments' in the experience of the transitionary threshold thinking of the pathway. These moments are not those of linear time, as they do not result in direct *instrumental* action. Here we attempt to free the sense of time from the calculative logic that its measurability inevitably sets up. This is the very 'outcome' of what Heidegger calls the '*other beginning*': a non-instrumental action in which there is a *transformation of our relationship to Being*. Here Being unveils what is concealed within itself, enabling new possibilities of engaging the imagination with the astonishing diversity of the world.

---

68. Heidegger calls this "reservedness" - *Verhaltenheit*
69. CP §13.

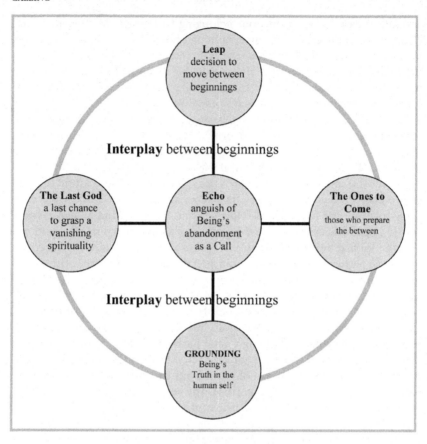

Here we follow an as yet unarticulated thought, a thinking of Being and its influence. It is an unfamiliar pathway that is not habitual, yet imparts the significance this experience returns to the sense of being human. The unifications give sustenance needed for this journey. However, their inherent unity does not lend itself to a sequential analysis of them. Not being a *system*, as for instance in the mode of the dot points of a strategy statement, they cannot be simply categorised and dissected, but are better approached via a manner of 'translation' I discussed earlier. Here I am not attempting to scrutinise the subtleties of Heidegger's philosophical project. Instead my aim is to illuminate those aspects that enable a pathway of homecoming that is a thoughtful movement of disruption and displacement in thinking from the 'already known' to an uncharted sphere that deepens the human experience. So here we begin a kind of 'promenade' amongst them. As they seem to suggest a between, or a threshold of the imagination, the 'beginnings' make for a good beginning.

Interplaying Beginnings and Horizons

The unification of the *"Interplay"* or *"Playing-forth"* gives us a particularly suitable point of entry into the nature of the *relationship* between the beginnings and its relevance to the thinking and experience of the truth that is other than empirical correctness. It calls attention to the 'playful' association between them. This play of overlapping spheres leaves its imprint on the circumstances within the threshold. Understanding the unfamiliar relationship that comprises the interplay between the beginnings enables their engagement in the transitionary and transformative pathway of the 'in-between'.

I have already suggested that there is a great deal of correspondence between the model of a New Philosophical Logic (p.54), the 'beginnings' and the notion of the threshold. We now know that Heidegger's 'beginnings' refer to two moments in the unfolding history of truth; the 'first' setting into motion the onset of metaphysics, the 'other' taking on board the primordial truth of Being itself. The 'end of the first beginning' is the moment when all the possibilities of metaphysics, representation and analytical logic seem to have been exhausted. Here I suggest that we may liken the 'end of the first beginning' with a 'home of familiarity', which has become inadequate (yet not to be discarded) for the task of retrieving a deeper experience of being human. Becoming aware of the necessity of that experience (the unification of the "Echo" explored in the next section) calls for a response; a 'Leap' that engages a movement away from such a 'home' of habit and instrumental logic, and is indicative of a lessened dependency on it. It recognises the impoverishment of the human soul and begins to know where to look for nourishment. Hence it becomes a crossing between a home that has become too confining, representing the restraints of uncritically and habitually held everyday, instrumental and metaphysical pre-conceptions, and an open space of new and unthought possibilities.

The threshold therefore is a place of passage that supports radical difference. Venturing it involves a stepping beyond the ego of the self to provide a clearing for a moment that conveys something of real consequence. Yet, the thinking of the threshold is in a region that is not fully 'outside' or 'beyond' the sphere of activity of the logo-centric metaphysics of presence. It is not wholly disconnected from the home of representation and practical objectivity. This means it is not isolated from the everyday as though in a trance. Instead, it is approached as a developmental and tentative pathway, an in-between, which builds upon an emerging awareness and a transforming conception of truth (from empirical correctness). Awareness moves beyond the usual sense of understanding; indeed the threshold embodies a far-reaching shift which resists our natural inclination for ordering, classifying and defining. As its pathway is sojourned, light is shed upon it, revealing a ground of human dwelling and illuminating a way of seeing, thinking and being that was previously held in darkness, beyond everyday thought.

To understand how the "beginnings" relate to the unifications it may be helpful to be reminded that Heidegger's crossing to the other beginning, as enacted in *Beiträge*, is a move away from representation to a different kind of truth, which for him is neither metaphysics nor ontology. It is a response to an encounter not of his making. As it evades the science of ontology, what matters now is Being itself. Here he seeks the union of truth and Being that is beyond observable facts and sequential time. The text of *Beiträge* is an outcome of his crossing between beginnings. It also corresponds with the shift in Heidegger's personal pathway of thinking, his *Denkwege*, as a "turning" within two idioms, between the thinking as characterised in *Being and Time* (*from* human beings as *Dasein towards* Being), and the thinking which apprehends the event (*Ereignis*) of, and out of, Being itself. In this pathway, the first beginning, as a historical event of Being, *first* awakens and brings about the flow of questioning. The thinking of the 'other beginning' is to do with those events where thinking is deeply influenced and transformed by the truth of Being itself. While the ontology of *Being and Time* overturned the hold of knowledge and reconnects with the other truth, Heidegger then, in the other beginning, attempts an originary thinking and enactment *from* this truth.

As an awesome surge of inquiring, demonstrated so well in *Being and Time*, the first beginning is still needed. Indeed, it is already indicative of the truth of Being. Yet, Heidegger insists that metaphysics runs into an unsurmountable boundary: it is not able to question Being in its grounding occurrence. This then calls for another way of thinking: the thoughtful attentive reflection Heidegger calls 'mindfulness'. Mindfulness in the crossing between beginnings is a thinking that becomes *aware of what is happening to it*. It comes to terms with the significance of *what it is*, that it receives from the 'essencing of Being' and from the 'turning' away from purely representative and calculative interpretations.

In seeking the ground of human dwelling, this crossing is not without its more immediate rewards. Heidegger, with allusions to the unique moments of unfamiliar insight that herald a homecoming, writes that 'the 'Interplay' is the "delight in alternately surpassing the beginnings in questioning"[70]. For him, in the crossing there is enchantment in his unfolding of the preparatory "*guiding*-question" of the ontology of *Being and Time* to the "grounding call" of the other beginning[71], which applies the guiding question to Being itself. The guiding-question was about the meaning of being, which laid the groundwork for seeking the truth of Being itself. While the earlier thinking of the ontological difference (the difference between Being and beings) was shaped and defined by its orientation towards beings, now the *simultaneity* and the *belonging together* of Being and

---

70. CP p.119.

71. CP p.165, §137.

beings is thought. For anything to belong together there must be a difference; or they would merely collapse into an indifferent sameness. Here their difference is not effaced but their unity is celebrated.

We have seen that the crossing over into the other beginning avoids metaphysical conceptions, yet the latter can only 'make sense' along with the first beginning of metaphysics. Indeed, I suggest it is *that* (the thinking of the first beginning) which ensures the other beginning does not become 'irrational'. For the other beginning to make sense, an understanding of rational logic is not to be discarded. Metaphysics is not to be destroyed; it is not to be defeated and left behind. It is rather to be "overcome". It is to be overcome in the sense of being transformed, renewed and raised to a new level of understanding[72]. The knowledge arising from the first beginning is not to be destroyed and replaced. The natural sciences are not to be stripped of their proper roles to enlighten and inspire. They will continue to provide much that is of benefit to human well-being. The social sciences still have to uncover a great deal to help us understand the culture. For instance, constructivist' theory has much to contribute to understand what makes society tick; how we organise ourselves and nature in ways and systems whose arrangements are too often obscured by its own language and standpoint. Undeniably, objective knowledge with its widely socialised intervention and its ease of transmission, storage and accumulation is the indispensable tool for the everyday regime. In our exploration, it is well to remember that, for those who wander the threshold of the void, on the border of the site of the event that puts us back in touch with Being itself, it is always fitting to be 'knowledgeable' in the first place. Unfortunately however, the orientation of constructivist and other rationalist thought above all considers the relation to Being to be *solely* within the dimension of information. Whatever is not predisposed to being categorised or interconnected in measurable ways does not really seem to exist for it. Therefore, it needs to consider the *origin* of any orientation of thought.

When thinking and writing about the *unthought* dimension in existence, certain metaphysical and analytical structures cannot, and need not, be avoided. Metaphysics itself is not an aberration. The first beginning is not to be destroyed, but metaphysics needs to be *ruptured* and transformed in order to retrieve the significance with which we are concerned here. The unthought and unspoken leaves its 'traces' in the veiled hints of the words of literature, which can be deepened in the thinking of the other beginning. Now we deal with the voiceless and nameless truth of Being in a way that attempts to surpass the need for representation. Yet, as Jean-Luc Nancy points out, as soon as one names the unnameable, even negatively such as the "non-presentable", or 'inexpressible', in a way it becomes a representation[73]. Nevertheless, here we are confronted with an unavoidable "tactical necessity", not a "totalising logic"[74]. Moving forwards on this

---

72. EP p.84.

73. FT p.229.

pathway of thinking would be impossible without such strategic means. Moreover, there is no reason why such thinking cannot maintain a certain distance between the limits of spoken language and the ineffable realm it attempts to describe; such a distance mirrors the very condition of human being-in-the-world. Articulated human language will always be lacking in its attempts to name and hear the language of Being. Nevertheless, to think of the truth of *Being as such* is possible, and such thinking can be recovered in the modern era. We need to better understand *what kind of thinking it is that thinks Being*, to *recognise* it when it disrupts everyday thinking, and to *enable the 'space'* that is required for it to occur.

Heidegger, in his dialogue on the "interplay" makes it clear that both beginnings are needed. Although he makes much of the experience of the pre-Socratic thinkers in which the first beginning arose, Heidegger does not advocate a return to such early Greek thinking[75]. We, who are but once, in this time and space and in this tradition with its particular history, cannot pretend its recreation in the impossible renewal of an ancient romanticised Greek world. That would involve a mistaken interpretation of Heidegger's first beginning, which is not to be reduced to a certain unchanging "original event" to which thinking might return. It would fail to see that beginning as a distinct *enactment of Being*. Because of the unique relationship between the beginnings he never refers to a "*second* beginning". The *other* beginning is one of a different quality that does not stand "over against" the first[76]. Instead there is a binding interrelationship and inter-articulation or inter-playing between them. He names the "*other* beginning" rather than a "*second* beginning", because "it must be the only other out of relation *to the only* one and first beginning"[77]. Or, to re-phrase this, the first beginning is a "*first* arising in relation to the other", while "the other can only occur through the first"[78]. Although it brings forth the other beginning, the *first* beginning can only uniquely be what it is *from within* the other beginning.

The "*interplay*" of the beginnings is a forgotten or dormant gift to every epoch, which does not exist in the past, or some point in linear time, but "lies in advance of what is to come"[79]. This means that as an *event of Being* the other beginning clearly relates back to the first beginning: "The style of thoughtful mindfulness in the crossing from one beginning to the other is also already determined by the allotment of the one beginning to the other beginning"[80]. Heidegger in *Parmenides* urges a transformation in thinking (a thinking that is 'under-way') "that brings us into the proximity of the *first*

---

74. As David Wood points out in TAH p.172.

75. PLT p.49.

76. See the various commentaries e.g. Alejandro Vallega, CCP p.55, CPI p.65, CPI p.67.

77. CP p.4.

78. CPI p.67.

79. P p.1.

beginning, in order to be closer to the beginning of the *approaching beginning*"[81]. In this perspective of non-linear, non-successive time, in its *'interplay'* the first beginning 'reaches ahead'. "The other beginning has to be realized totally from within *Being* as *Ereignis* and from the essence of its truth"[82]. All this signifies that the beginnings are *unique* occurrences of Being that efface sequential time. The event Heidegger calls *Ereignis* already happens in the first beginning, but as thinking 'overflows' its limit and we approach the other beginning we become *aware* of its unique character. The above descriptions define the *manner* of transition, of "playing-forth", between beginnings. When thinking is transformed in the overflowing of the imagination the two beginnings are brought nearer and their unity is sensed.

A primary feature of the first beginning is the originary comportment of awe and wonder. This is not the awe of the terror of the unknown, or the hypothesising of why, what and how. It is simply the beholding of 'how things come to be'. This is not 'lost' in the other beginning, quite the reverse: it is *grasped* as something essential that grounds human beings. Without it we can only become ever more technological, mechanistic and calculating in our ways of thinking. The first beginning is not to be brought about through the repetition of an unchanging origin, for instance as in the re-embodiments of religious ritual. It is instead to be 'opened up' into its diverse possibilities, whereby the other truth may be experienced *and* questioned. The primordial thinking of the first beginning arising out of Being itself can be experienced uniquely today. However, for this we need the 'light' of the other beginning, to illuminate what is uncovered in the first beginning of the western metaphysical tradition.

Therefore, the important point to be made about the beginnings is that, although they have distinct characteristics, they, like their fugal unifications, cannot be neatly categorised and should not be understood as defined points in history. The movement in thinking between these beginnings is not simply chronological, as if it were an issue of continuity and discontinuity between eras in history. In originary thinking we do not simply 'move' from one to another, but we can, in a way, sojourn their 'in-between'; thereby facilitating the threshold. 'Adding' a third, 'meditative' kind of thinking, or mindfulness, that recalls origins overcomes the linear progression from the early Greek thinking to utilitarian objectivity. Arising out of the non-sequential temporal nature of being, there is an *encounter* between them, whereby they "play-forth". Because the threshold has its origins in these beginnings, its pathway is also not sequential in terms of space and time and does not have an exclusively linear progressive structure. The unification of "*interplay*" between beginnings ensures that the threshold cannot be restrained to a fixed space/time framework. Yet the

---

80. CP p.4.

81. P p.139.

82. CP p.41.

time and space of the threshold is a means of access between the familiar home of the everyday, and an open but untrodden space beyond. It is a pathway of thinking that involves an *interaction* of time and space, which characterises the various 'moments' in its experience.

When the other beginning comes into play we enter "into a confrontation and dialogue with the (first) beginning"[83]. This is not a confrontation that destroys or divides. Being in the threshold does not exhibit a Nietzschean characteristic of being torn between two worlds: a place of combat between the treachery of the human heart and nature. Yet, Heidegger insists that the crossing involves a degree of 'strife', a struggle to let a deepening of understanding emerge that builds the necessary unity between the ground of human dwelling and the ground of human building; between the world of human activity and the fundamental human essence. In the 'strife' of the crossing the familiar and the habitual is ruptured. This rupture has an astonishing outcome: it rips apart the everyday shackles of instrumental thought to create an open space. The turmoil of the threshold does not collapse into chaos. The strange temporality and spatiality of this openness lets in unseen possibilities; possibilities emerging from the other truth that have the potential to authentically ground human experience. As we participate in its interplay an event of consequence is accomplished. However, it can never be completed in the sense of being finished with it. It does not involve the kind of consumer experience of the everyday which leaves behind nothing of significance. As beginnings, they *initiate* something and here these indicate a 'being-underway' into a new and different relationship to beings, the being-ness or essence of beings and to Being itself. In this in-between where meaning arises, humans find the foundation for genuinely 'building' in the concrete world of the everyday. Here we can "proceed to that which is erectable and preservable" within the creative domain where beings are disclosed in their essence[84].

I shall return to ideas of openness later, but for now we pause at the limit of the everyday with its corresponding boundary or 'end' of representative metaphysics, so we may recognise the implications of venturing such a horizon.

Limits

We have seen that the 'end of the first beginning' may be linked with a 'home' of familiarity, which has become inadequate if it is to retrieve the deeper experience of Being. The crossing then is a *'between-like' transitionary thinking*, in the sense of being *'underway'*, past a kind of a 'boundary', where thinking flows *beyond* everyday ways of thinking. In doing so, the imagination enters a *region of thinking* where we are not fully disconnected from

---

83. P p.166.

84. CP p.314.

everyday calculative, representational and metaphysical thought, but nevertheless experience a *rupture* of the habitual as the limits of valuation, calculation and objective determination are exceeded. Heidegger's own 'crossing over' or 'leap into the between' is over a kind of *limit*, where the first beginning comes to an 'end'. Again, we need to understand the ambiguous nature of 'end'. Is it mere cessation, where something can go no further and ceases to be? To be sure, this 'end' implies that the inadequacy of traditional perceptions and propositions seems to close off further possibilities. Thereby, religion has played out all its possibilities and, unless one's emotional commitment is unbreakable, the only option appears to be the total surrender to utilitarian instrumentalism.

However, Heidegger's work on 'overcoming' metaphysics shows that when the 'other beginning' comes into play, the end of the first beginning becomes a margin from which something *new* is able to emerge. The end becomes more like a *horizon*, one from which "*something begins to truly show itself*"[85] and "*whereby beings first come to be*"[86]. A place then emerges that, to the poetic imagination, indeed reveals as a between; between gods and mortals, between the dark earth that seems to withdraw 'the whole' and the light of Being. For Heidegger, this engagement with the other beginning in turn retrieves the inceptive origin of the first beginning; it saves what was truly significant in it. At the 'end of the first beginning' there is then a sense of a boundary from which beings are again able to be revealed more fundamentally; i.e. in terms of their essential nature, which retains *concealment* as its precondition. For Heidegger, concealment lies at the basis of revealment; it belongs to Being like shadow to light. In concealment, what is withdrawn is thereby sheltered; it is thereby a kind of *reserve* that awaits the human capacity for perceiving and revealing. Therefore, the withdrawal is not a negative closure, but provides the necessary reserve that *provides for*, and *protects* emergence, as an unfolding from the inexhaustibility of truth, without which poetic imagination and human creativity is impossible.

To approach this reserve is to be underway to the 'something more' implicated in the human need for meaning and expression. It requires the 'risk' of venturing outside familiar territory to the 'strife' of the threshold; hence it is not for the fainthearted. However, this also cannot occur if we place ourselves at the centre of all activity; the illusion, enforced daily in the restless discourse of the culture, that it all depends on us. In the contemporary world it often seems that absolute and compulsively amassed knowledge, which 'explains' everything about us and around us, is all that matters. The rational mind has placed an absolute limit around itself. Yet, fortunately the imagination is not so readily contained. Thinking transforms when it exceeds the everyday; beyond the limit to where it is translucent to the pure open beyond. It recognises that we can never know the full

---

85.PLT p.152.

86.IM p.63.

story, i.e. 'the Whole'; there is always something more that is unthought and unknowable. This is not because we do not yet have the tools that are adequate to the task of knowing 'everything', but rather that "unknowability is a feature of reality, i.e. of the interaction and interdependence of our Selves and Nature"[87]. Heidegger understands that true philosophy holds something deeply essential that unifies it and which is expressed by all great thinkers. This most fundamental ground is "so essential and rich that no one individual can exhaust it", but rather unifies all such elemental thought[88]. 'The whole' must always be a matter for thinking; without that the culture becomes too exclusively instrumental, too mechanistic, too lacking in poetic imagination. Yet, at the same time this questioning is only an uneasy 'crossing'; a struggle with the immensity and mystery of this truth.

Mystery, not to be confused with fashionable forms of hastily and effortlessly acquired mysticism, is mostly absent in modern life, where the culture demands the illusion of certainty and scorns doubt. Yet, it is an essential need, one whose absence lurks at the groundlessness of contemporary quandaries. Instead, if the mystery is encountered at the limit of objective everydayness it becomes that lure within truth itself that draws us to the truth. It does not seek to find *the answer*; one that would conclude questioning as thinking has reached some perceived exalted state. It rather *deepens* the need for venturing the mystery of this truth, which is withheld from view beyond theory and accounts and cannot be known objectively. Something truly new (i.e. unique) may yet emerge from the barren cinders of metaphysical representation, so brutally torched by the high priests of science, and so carelessly disposed of by the disciples of totalising logic and calculative evaluation.

So, when the first beginning comes to an 'end', it has attained a limit and yet it does not die away. Limits define and give stability, but they also bring into play the undefinable; that which is lacking or absent. Here, we are addressing something that is unique to the nature of the interplay of the beginnings. The limit of the first beginning demonstrates the impossibility of resolving a certain line of questioning. As discussed, to go further, the constraints of pure logic have to be left behind, not to be substituted by something 'illogical', but by a 'reflection' that is sensitive (open) to what might be happening on the 'other side' of the limit. Thinking at the limit must '*overflow*' if it is to continue and, moreover, if it is to enable a transformed way of thinking that we can *take back* to the everyday, so that life in the familiar domain may be practised in a more authentic manner.

Heidegger writes about this "overflowing" of limits. It is not a matter of "too much quantity"[89], but rather involves the self-withdrawal of a calculative, instrumental and representational attitude, thereby allowing a growth

---

87.F. Fisher, op. cit., p.268.

88.*Der Spiegel* Interview.

89.CP §131.

in thoughtfulness that clears the open 'space' necessary for an event of significance to occur. In this ineffable space the imagination overflows from straightforward experience. Then form and figurations, presented for human imagination, reveal their being from out of their very ground. What previously were merely outward appearances are now unlimited from the constraints of representation, and able to overflow the limits placed upon them. The light of Being, as a gift of primordial truth, then bursts forth out from 'the pure open', the 'other side' of the threshold, whereby things are sensed as something more than objective representation. The threshold is the 'in-between' realm that is illuminated by this light.

Therefore when we involve ourself in the crossing of the threshold a region of transitional and transformative thinking is traversed, across the boundary of the constrained homeliness and logic of everyday familiarity. Here thinking moves towards a particular kind of worldly disclosure, where imagination is offered a less defined and more elemental freedom. Everyday experience undergoes the effacing of its usual subjective character. Although the imagination spills over everyday limits, our response to the excess still respects these boundaries. It is not as though we become superhuman; indeed this is a region of humility where human arrogance is shown to be based on an illusion. However in the threshold these limits are not the stifling constraints of the instrumental attitude, where the surplus hardens into mechanistic objectification. Thinking is thereby also less influenced by the changeability of vague feelings. By going beyond these the threshold is engaged as part of the journey of 'coming to be at home', a pathway where everyday awareness overflows, accomplishing the richness of human potential and merit for earthly dwelling.

A Reverberation from the Gods

The 'fugue' that Heidegger names 'the echo' (*Der Anklang*) gives us a supportive direction to grasp the unfamiliar moment that interrupts the everyday way of thinking. It is a call for homecoming. The echo is a reverberation from the very ground of our being. It originates from our primary foundation: the site of the source, the void or abyss, where the gods rest in its stillness. The notion of the 'gods' will become clearer as we continue, but for now it is important to understand that for Heidegger these do not refer to entities, but rather *occurrences of Being*. These are moments of deep significance that concern existence itself.

As discussed, Heidegger perceives one of these occurrences as the first moment of the history of Being; the 'first beginning' of primordial Greek thinking. In his reflections on their writings he reaches deeply into this 'earthly' essence of Being as *physis*: i.e. as 'Nature'. For him, the modern term 'nature' is too ambiguous, distorted and overburdened with a bewildering range of derived meanings. The diversity of disciplines, such as environmental science, eco-philosophy/psychology and environmental ethics, bears witness to a concept that is susceptible to a wide spectrum of

69

interpretations. Yet, for Heidegger, none of these reflect this original full-ness of the Greek concept that resonates in the expression "*physis*", of which 'un-concealment' is its fundamental character. He regards the early Greek notion and experience of Nature as 'emergence', i.e. the 'bursting forth' of itself in *fidelity* to its vocation for revealing and concealing[90]. Then, in this elemental experience of being, the gods were first able to be placed. This for him *is* 'Being'; not simply all of nature, but rather the gift or the *gesture* of unveiling something which is concealed within it. The primordial Greek thinking concerned itself with Nature's originary truth[91]. It was a realm where the gods seemed unimpeded to come to pass. However, from Plato onwards, Nature as *physis* inevitably became taken-for-granted presence, from which, inevitably, rational objectivity arose; thus completing (bringing to an end) the Greek thought of being as appear-ing. The Greeks were not able to *keep* the gods; their vulnerability to an ex-cess of form became too much for them. This subsequently led to the usurping of the original essence of truth. Representation then became the only prospect for the inevitable ascent of objective knowledge, whereby in the 'flight of the gods' the experience that grounds *all* experience withdrew from the human grasp. Unable to 'release' or 'open up' ourselves to the self-emergence of the world, mere appearance has become the sole and decisive interpretation of being. The gift of unveiling has merely become 'presence'; as the obvious feature of how we see nature, ready for 'capture' by our digit-al cameras or as a stockpile for exploitation. No longer able to let this un-veiling come about, this becomes a *lack* whereby we are abandoned by Be-ing as appearing and as constancy. Yet, despite this abandonment Being as *physis* maintains itself, as it is always faithful to its relationship with human thought, if only we would become attentive to it and let it come to pass. By listening to 'the echo' of the fleeing gods, rational objectivity can be saved from becoming self-enclosed within instrumental frameworks.

The echo is consistent with the need of human beings for something more than objective, utilitarian and instrumental logic. It is a recollection that comes from Being itself. That means that it is an experience that is fundamental to all experience and originates from the silence of the void. Perhaps it is as Lao Tzu stated much earlier, "In the clarity of a still and open mind, the truth will be revealed". There needs to be this silence, for something new and unique to be able to enter the space and time of the moment. This puts us in a readiness for hearing and preserving another kind of truth. For the ground of human dwelling to unfold, the deep still-ness, as the truth of Being itself, must be sheltered in the world. This readi-ness keeps oneself open to the arrival and passing of gods. The stillness of the echo maintains the capacity for the safeguarding and sheltering of that what must not be objectified and consumed. It preserves the capacity for discernment of what truly matters and what does not.

90."the emergent placing-itself-forth-into-the-limit" (IM p.63).

91.CP §20-23.

In western philosophy the call of the gods was first given voice in the emerging pre-metaphysical thinking of its early Greek formation. Since then, as instrumental perception takes hold, the echo becomes a recollection of this 'call', involving a kind of anguish, like a trace of something that seems to be missing. This indeterminate sense of loss or vague unease is a silent summonsing for departing for the crossing of the threshold and 'becoming at home'. Today it reminds us of the need to recognise the anguish of the poverty of spirit arising out of modernity's loss of a sense of existence itself, or 'being as such'. The essence and ground of human existence are constant gifts. The echo reverberating out of the call of Being tells us that the ever dynamic emerging of Nature endures. The reverberation persists, despite the contemporary losing sight of the ground of human beings, since it is founded in the *need* that comes about in the abandonment of Being. However its resonance has been lost under the ever-growing and congealing pile of materialist hubris, or forgotten in the restless quest for diversions. Although few respond to the call of Being in the epoch of the overwhelming hold of technological thinking (Heidegger calls this *technicity*), the echo endures as a beckoning for homecoming to the ground of human dwelling. In the human soul there persists the origin of a deep stillness that is able to discern the echo of the gods, despite the noise of striving after things and amusement, and the hold of calculative thinking. It may be ignored or forgotten, but it does not vanish. *Something*, the need for 'something more', *remains* in memory albeit hidden and unexperienced. When this need is suppressed, it re-emerges in destructive forms such as obsessive consumerism, narcissistic individualism, depression and nihilism.

The primordial beholding of emerging nature may be recovered in the crossing of the threshold. There, the reverberation of the need may be heard again and sheltered. The echo is perceived when the *loss* of a meaningful sense of existence is grasped. When this loss is no longer abandoned to ostensible irrelevancy, there is an affirmation that losing sight of Being has its costs and something has to change: it is an experience that is *transformative*. The possibility of a moment of sensitivity to this event cannot be annihilated, but may be taken up as a lasting awareness of a hint; one that touches on existence itself and reminds us that the angst of the void is not to be evaded. We are rather to come to terms with it, i.e. 'become at home' in it. Both the angst and its affirmative companion 'wonder' belong to Being as such. They therefore maintain a non-epochal, non-historical aspect that is beyond the reach of culture and tradition. Their echo is beyond the fleeting dreams of redemptive religious faith. It is what remains of Being; what is still available despite the distractions and the impoverishment of the era. The reverberation is indeed a kind of offering, an *excess* of Being as such. Such a surplus may be "more difficult to bear than any loss"; as the "greatest need" one *must* do something with it[92]. As this need is taken hold of, we are compelled to retrieve the deeper experience of "Being's

---

92. BQP p.133.

essencing" (*Wesung des Seyns*). Then imagination overflows into the turbulence of the crossing of the threshold. The mystery of the excess is not to be bottled up, nor is it to be reduced, or 'explained' into oblivion. In fidelity it is to be responded to. It is to be *liberated* in the open clearing of the threshold, as '*the poem of Being*' that names Nature in its essence.

This then becomes the other beginning, where the difference between simple presence and self-emergence or dynamic revealing, becomes a matter for thought. The written word cannot become a poem unless it impels us to enter the threshold of being. In the first beginning the 'poem of Being' was simply sensed with its infinite possibilities; however, subsequently the poem was interrupted by representational metaphysics. This does not mean this interruption is to be deplored; without it, innovation, the adventure of seeking and gathering of knowledge and the fruits of science and technology, could not emerge. The bond of the poem to emerging Nature, or Being as *physis*, is not meant to enchain human orientation to a single way of being, but rather is to endure as the underlying ground for its interpretation. Yet, the western configuration of thought is out of balance because it does not tolerate a radical inverting disruption by the poetic theme of Being. The calculative usurping of Being's poem has become a project of mastery, whereby its enchantment and faithfulness is supplanted by a technological way of thinking that has become totalising.

The experience of the withdrawal of Being is not all bad news. As something that is structured beyond human control, it offers the insight that we are in an essential relationship with Being. Heidegger is fond of quoting the lines of Hölderlin's poem 'Homecoming': "But where the danger is, there also grows our saving grace"[93]. The essence of technology can reveal a way of being that would otherwise remain hidden. He believes that in the 'granting', i.e. in the manner in which it is able to *reveal* a need, lays the saving power of technological and representational thinking. When grasped as the "distress of a lack of distress" we have awoken from our slumbers. As already suggested, a pre-requisite for venturing the crossing of the threshold is the *realisation* of the seductive powers of the mastery and ambition of calculative and technological thinking. The echo accomplishes that role and sustains the human belonging to being.

As noted, the echo, or *hint*, reverberates and originates from the very ground of our being. This is our primary foundation: the silent site of the source, the void or abyss. It is the neglected ground of human dwelling that 'calls' for its retrieval in an enduring way. For this to come to pass, human beings need to learn to live with its stillness. This region of silence is not simply a lack of noise, but rather is the absence of "idle chatter" allowing the return of meditative thinking. Detached from the hold of instrumental and technological thinking and released from the need for the reassurance which the everyday always attempts to provide, we sense an inner stillness that turns towards what stills and are no longer in fear of it. The

---

93.EHP p.40, QCT p.35.

restlessness of the everyday is converted into '*wandering*'. We become 'wanderers' whose 'innermost heart' or 'soul' (I return to this later) is attuned to the "deep stillness" of the peace of the foundational site[94]. Seekers of the ground of human dwelling are ready to 'catch' what is being directed to them: namely the call or echo of the ground itself[95]. In grasping this domain with its otherness unto human experience a region is entered that cannot be possessed. This is a sphere where the ego is effaced and the otherness of Being and its unfamiliar truth is allowed to speak in a way that was not there to begin with. It becomes then a kind of re-enactment of a primordial experience unique to human beings.

Moving into the stillness that comes to pass from the existential void brings us nearer to the mystery of the truth of Being. Yet we remain distant; we cannot find it like one finds a lost object. The ground has an indeterminate abyssal character; it is self-sheltering and concealing and therefore it can only 'hint'. We may recall the cryptic comment of Heraclitus: "it does not say and it does not hide, it *hints*"[96]. This hint is perceived in the echo, where the withdrawal or forgetting of Being is experienced as a deep need. As it simply hints, the claim that the truth of Being makes on us does not strike with great force, nor does it offer any assurances. Although it does not explicitly announce itself, it is essential for creativity to emerge and flourish. Those with enough poetic imagination feel this need when objective discourse just does not seem to tell us enough about life and something deeper needs to be uncovered. When perceived in its resonance as a moment of significance, the *hint* of Being points towards the distinctiveness of a threshold site. The creative person recognises something of consequence here that cries out to be articulated. It is a place of consequence because it enables the opening up of what is sheltered and concealed. 'Hinting' leads into something that is to be understood, i.e. into the dimension within which one understands. This is not only 'alluding to something', but also *guiding* into making contextual sense and making it '*meaningful*'.

So, the call of Being remains unspoken yet is articulated in a creative response. The extraordinary thing is that it is existence itself (Being as such) that hints. It beckons, and is remembered and retrieved with possibilities of richness, of fruitfulness, or of "ripeness", where "Being itself comes to it fullness"[97]. "Fullness" indicates the readiness for the gifting of its fruit. The hint indicates a gift that is ready but 'not yet' given; it announces the 'readiness' for another unification, "the Leap" into the thinking the 'between'. The hint arises out of Being's character of self-withdrawal and concealment, which never irrevocably buries itself. It brings into the open

---

94. CP §13.

95. See CPI p.33.

96. Cited in H. Arendt, *The Life of the Mind*, op. cit., p.210.

97. CP p.288, David Crownfield in CCP p.224.

the kind of saying that "speaks from the truth of Being" itself rather than from "logic"[98]. Such 'saying' is thoughtful, mediatory and poetic; its manner of expressing gathers, understands, protects and sustains the silence that is inherent in the call of Being. Here, near to one's ground, the truth of Being is sheltered in beings, both through its disclosure in beings *and* in its withdrawal back into indeterminacy. The seeker of this truth knows that something inexhaustible has been held back 'in reserve', which remains 'near'. The truly creative person knows there is always more of this bountiful reserve from which to draw to his or her creative inspiration.

"The echo" then is the reverberation of the call of Being. It may be perceived as an unspoken cry of a hidden and forgotten self-understanding. The experience of the hint of the "silent voice of Being" is not fixable, calculable, measurable or definable. Michel Haar refers to this silent voice as "the speech of Being that no-one has spoken". It is the unwritten and unwriteable "Greatest Text"... the "unspoken gathering of *logos*"[99]. This refers to a primordial language that precedes human speech, not as some 'primitive' pre-human language, but rather as a trace of the mystery that is Being itself. It is a source that 'speaks' out of its truth and from which articulated language then comes into view. So, this voice is unlike the vociferous proclamation of a prophet, to be rejected or received in faith. 'Listening' to it cannot be akin to taking a transcript from the sacred, like Moses writing down the words of the Lord's covenant[100]. Rather than a teacher/pupil or master/servant relationship, here we encounter a wholly other union between the human imagination and the unspoken voice. It sustains the mystery that 'draws in' the thinking of another beginning which embraces an elemental truth. It entails what Haar describes as "hesitation between sound and sense"[101]; an instance of wavering between objective perceiving and sensing an otherness beyond words.

Heidegger famously called language "the house of Being". We return to this again later, but for now we can say that, although the essence of language originates in the essence of Being[102], it is mortal human beings who are needed to articulate language. *Human beings are needed for the revelation, protection and configuring of Being.* It is not just for the educated elite. Although few respond in the modern world, the language of Being speaks in all, as it is "insensitive to acquired knowledge". The echo summonses one to a different kind of truth. Upon hearing this primordial language, human beings can respond and express its unspoken word. This is how Being is always "*underway*" to language".

---

98. CP p.55.

99. SE p.151.

100. Exodus 34:27-8.

101. SE p.112.

102. Cf. CP §276.

The void, the pure open void of Being itself, reaches out for the 'soul' of human beings, whereby they are thrown into the turbulence of the threshold. Although arising in response to stillness, the threshold is not tranquil. Uprooted from our slumber, the "greatest need" of the modern era is recognised: the reclamation of the most innermost self, or true "self-*being*"[103] of human beings. Thereby, despite its origin in the silence of the ground, the echo becomes a shattering call that disrupts the familiar and the habitual. It demolishes the adequacy of the safe haven of familiarity which may be seen for what it is: a home without substance; one that is only provisional and without essence, borrowed and derivative, not arising from an authentic ground. When the usual human relationship to objects and structures that only *appear* to protect us is understood as being inadequate to the human essence, we are left vulnerable to the void; the abyss of which we need to make sense. The echo provokes a thoughtful response. The response to the experience of the abandonment of Being is now to partake in this 'strife'. This need beckons an unfamiliar kind of 'decision' that Heidegger calls "the Leap"; the unification of *Der Sprung*.

A Leap in Faith?

We have seen that the space of the threshold is a place of exile, in terms of being out of familiarity. It is an experience that can be ambiguous and disturbing; its abyssal nature may be overwhelming. Therefore, in human homelessness, it is usually disparaged or avoided. Yet, as the existential void lingers in the background, we witness impossible attempts at its 'filling in', such as via the distractions of materialism, the consumption of events and experiences and the seeking of refuge in forms of religion that come packaged complete with a warranty of eternal salvation.

However, to become at home we are not to flee the angst of the void, but rather we are to depart and wander towards it. To be on the way to the source that is pure being and offers a transformed freedom does not mean that we thereby *separate* ourselves from the home just departed; we are not withdrawing from the everyday world as into some enclosed religious order. Rather in the essential leaving we are at last exercising our *fidelity to* the world. The beginnings are then gathered into interplaying unity. The departure is essential if we are to retrieve our humanity and render justice to it. We "leave the familiar shores of our experience of truth in order to venture the unclear waters of another". Only then are we able to "touch on the ontological soil on which the experience of the loss of Being has grown"[104].

When one becomes a wanderer who seeks the truth of Being, he or she throws aside everything familiar. While previously captivated in a pre-

---

103.ET p.154, CP §19.

104.TWH p.175.

defined world of objects (and set over against these in the desire for mastery), the wanderer expects nothing from beings immediately (in the sense of utility). Sojourning this pathway of the crossing of the threshold is accomplished in a stumbling, modest and uncertain manner; not a wilful moment of the ego. Letting go of the autonomous self makes "the Leap" necessary, as the moment of unguided decision which is not entirely our own. It is now 'prompted' by an event of Being. Yet, while keeping open such a moment is now not so much a matter of the 'will', to respond to the mystery one still needs to retain a certain measured release of understanding, care and restraint.

As the everyday way of thinking seems so obvious and absolute, hearing and responding to the silent call seems at odds with our natural instincts. It necessitates a movement from the striving for things to *no-thing*. As a passage from the familiar to the unfamiliar, the threshold of thinking is approached hesitatingly; one cannot be certain of anything here. By implication, it calls everything into question, casting the human being out of a habitual ground, opening before us the mystery of existence. The hesitation is intrinsic to the transitory thinking between beginnings, where in the crossing of the threshold we are not fully disconnected from the first beginning, and at the same time in the 'not yet' of the other beginning, which is *hinted* at. We are only *beginning* to sense its awesome truth.

Those willing to risk being out of the comfort zone to be underway toward the source, the ground of Being, may 'flee' the used up and abused world, the worn-out home of human misuse. The need for a timely flight into the solitude and subtle beauty of the natural world is not the outcome of a failure to adapt to the 'reality' of the mob. Perhaps it comes about by a disillusionment with the illusion of our times. As a response of thoughtfulness it may well be a search for a different truth, often scorned in the assertive culture that deludes itself that it has everything under control. To become aware of the greatest need requires a crisis, perhaps like the one we face today, so that we truly *experience* the despair of realising that we have deceived ourselves. This disillusionment is the first step for the return of *humility*, which for Heidegger is the prepared soil for authentic growth, poetic imagination and creativity. Today, talk of creativity instantly evokes images of cool young geeks in advertising agencies or design departments busying themselves in front of giant monitors, being helped along by the latest 'intelligent' software to produce fleeting experiences, or soon to be superseded novelty. Unfortunately, some activities that go under the guise of art, are also little more than contrivances of gimmickry. They are devices that merely aim to break into the short-lived attention span of the 'time-poor'; to be observed momentarily, described as 'amazing', and then discarded. These are not experiences that transform and enrich, leaving something of consequence. Such ploys do not fulfil art's mediatory role of illuminating a path between representation and truth. Notions of creativity need to be rescued from unthinking and degraded use. Genuine creativity plays its proper role by unveiling meaning by way of a struggle, a gentle wresting from concealment in the many 'between' regions of life's 'strife'.

The creative response does not happen without a corresponding transformation in the human psyche. It is a journey for which we need to divest ourselves of the baggage of power and the idea that there is no limit to human cleverness. If we think of ourselves as philosophers or scientists, we need to remind ourselves that we shall never arrive at absolute truth or a Theory of Everything! The truly creative person is humble in the face of human fragility and transience; thereby the capacity for awe and reverence grows. The grounding nature of the threshold is within a pathway where human beings can authentically dwell in what is one's own. Authenticity is not some elitist state of being, as the assertive nature of its Greek root may suggest. This is where the German *Eigentlich* seems to be nearer to the conception of being true or *proper* to one's self; aware of one's essential nature and that of all beings. Michael Leunig observes that the true self of modern man seems to be lost or has "gone secretly mad with fear and exhaustion and is too weak and frightened to emerge" to be faithful to itself[105]. Sheltering and nourishing this healing faithfulness is missing in the disenchantment of modernity. This faithfulness is quite contrary to the unconditional faith of religion. The reverence of which we are capable is not for a supreme power, but for the primordial mystery of life.

If Heidegger's fundamental thinking has the potential to reawaken the question of faith, it would be a transformed concept that is without dogma, one that keeps open the 'reveal-ability' of the divine, or the *possibility* of revelation[106]. It would be a faith that is prior to the determinate faith of positive religion, a sense of 'trust in truth' that leaves open fresh possibilities, and shelters the mystery of the unknowable. Heidegger's (like Nietzsche's before him) destruction of the metaphysical representational God does not eradicate the offering of 'the Other' that calls for a response. The response is an act of faith, one that is neither objectifying nor subjectifying and measures neither knowledge nor certainty. It may express itself in writing, in art, in a song, a sound, a touch or a dance, any gesture that requires the courage of becoming an accomplice in the event that invokes 'the strange'. It is not my aim to argue for a kind of 'secularised or private religion', with its implied need for prescription and rule. Here I simply attempt to reveal a way of being that is an authentically human and deeply significant. It may even be described as 'spiritual', as a transformed notion. My hope is that this book demonstrates that meaningful human existence can be understood and experienced deeply; that it is possible to stand in awe and wonder at the beauty, abyssal dread and the complexity of life in all its manifestations, without the support of religious belief. Anyone who

---

105.M. Leunig, op.cit., 'Hello, welcome to our drought'.

106.Derrida, Jacques. The Gift of Death, Chicago: University of Chicago Press, 1995, p.49. This very readable (for Derrida!) volume contains considerations on death, dying, moral and ethical responsibility, religion and theology.

seeks meaning beyond a purely rationalist and instrumental worldview, who has grasped the need to include questions about existence as such and knows that a response is called for, can be considered as exercising a fundamental spirituality.

The Leap is the 'departure-event'; a response of *fidelity* to the offering of 'the Other'. This faithfulness is both to the 'home' left behind, and the act of the crossing. In a later section we shall see that this fidelity maintains the 'gods' once their return has been ensured by the intervention of "the Ones to Come". For Heidegger, these are the poets and thinkers that know how to distinguish the alluring but narcissistic stories of legend from their underlying sacred message. Yet, these are also all of us, who as the keepers and guardians of the primordial event hold sufficient poetic imagination for the coming and passing of gods. We cannot 'reach' beyond the crossing; instead, that which is beyond *reaches us*. We are simply to accept our proper function: a friend to the event of consequence who celebrates its treasures; a companion who in such fidelity attempts to unfold its truth, as a foundational and sustainable offering to human thought.

In the displacement of the unique event we are 'alone'. The transformative dislodgment can only occur in one's personal pathway of thinking. This is not some miserable individualism as 'meaning' can be retrieved, but not possessed by me. The unifying ground of Being is the source shared between all beings, yet it is only 'I' that can 'hear' the "echo" and "leap" into the pathway of the crossing, and think the other beginning. Therefore, this may come about in the times when solitude helps provide a kind of prepared space; where something other than own voices can be heard, opening us up to that which is given. It seems that the great thinkers, poets etc. often got their inspiration in solitude, on solitary walks in the countryside. Of course, amongst the rest of us there are many who also sense a greater clarity or a renewed creativity on such excursions. Within the forest, the remote shore, the silence of the desert, the confusion of the everyday fades, allowing the unformed earth to self-reveal as a primal happening. Perhaps the absence of distractions, away from the individual 'direction' of other human beings, and being amidst less aggressive beings, helps to provide the 'open space' and stillness needed to hear the reverberation of Being.

In the threshold, beings are 'allowed to be'. Heidegger's description of the comportment of "letting-be"[107] involves letting things come to presence in their essence. Being released from the hold of the desire to cause some sort of effect and to physically or conceptually possess and control facilitates the calmness, or stillness, that prepares one for "ontological openness to the no-thingness of Being"[108].The 'leap' involves this letting-be,

---

107.*Gelassenheit* - a term Heidegger borrowed from Meister Eckhardt.

108.Leslie Paul Thiele, *Timely Meditations – Martin Heidegger and Postmodern Politics,* Princeton: Princeton University press, 1995, p.75. See particularly the excellent discussion on letting *(lassen)* and the nuances in the way Heidegger uses all its derivatives in Richard Rojcewicz, *The Gods and Technology –A reading of Heidegger,* Albany: State University of New York Press, 2006. op. cit., p.32ff.

whereby human beings anticipate the possibility of originary freedom; a freedom that has been revealed from Being rather than the utilitarian freedom of supermarket choice. The letting-be of beings is utterly contrary to indifference, which is the absence of care. The letting go of the desire to control does not imply a state of complacency, 'fatalism' or a passive withdrawal from the problems of the world. To anticipate the disclosure of Being is to *expect*, what Heidegger calls, the god. To 'hear' the 'echo of the gods' involves a "waiting that leaves *open* what we are waiting for", and being able to wait, "even for a lifetime"[109]. Rather than this being a condition of passivism, it means that Being's revealing needs a continuing human openness; one that endures and anticipates. It implies a human 'attitude' faithful to its essence and the relationship with the ground of Being. Its thoughtful questioning does not demand completion. Yet, in the impatient and restless age of the technological attitude it is presumed that this kind of waiting is not necessary; indeed as there is nothing that should not be on call, it is not to be tolerated.

The movement into the threshold and letting beings be engages the action of '*touching*'. What it touches on beyond the sphere of everyday consciousness is Being itself. By touching on the very sense of the dignity of existence, it does not absorb it as a representation. The belonging of humans to Being is affirmed and enacted here; the relationship becomes 'intimate'. Decentring the human being awakens us to an *originary ethics* that touches the being of all beings. This *grounding* for responsibility recalls the original Greek experience of *ethos*, which ponders and sustains the "abode and essence of man"[110]. From this originary ground of ethics, society's 'values' are derived, which may be applied, disputed, reformulated and adapted according to the worldviews of the culture and the depth of thinking it sanctions. Ethics then becomes like an ecosystem founded on the ground of *ethos*. This proto-ethical soil must be preserved if the growth and integrity of practical ethics is to be sustained and nurtured. Fidelity to the awakening of this sense of existence will be proper to a sense of 'shared being'. Letting-be is a response of fidelity to beings; it shelters their immutable bond to the truth of Being, so that they can show themselves as the beings they genuinely are[111]. Then, in the renewed awareness of the threshold Heidegger's 'divinities' may be perceived as what bring something to life: in their emerging self-revealing and concealing they *are* what glow in it. Not as a mere sense that we might responds to as 'pleasant', 'disagreeable', etc., but as a sense of being, or being as sense[112]. This is not a sense of transcendence whereby experience surpasses existence and entities towards a

---

109. M. Heidegger, *Discourse on Thinking,* trans. J. Anderson and E. Freund, New York: Harper & Row, 1966, p.68; IM p.221.

110. ZS p.217, note.

111. M p.86.

112. Cf. FT p.187.

'cosmic beyond', i.e. a mystical experience. Instead here transcendence itself enables a kind of *communication* with the beyond, yet without losing the desire and the ability to make rational sense. Nonetheless, the mystical experience in not unrelated to transcendence; indeed it experiences that which 'brings about' the experience, but does not recognise the source itself. In reflection, the deeper significance of the experience of transcendence gathers all beings towards that origin to which we already properly belong.

The site of the threshold is the 'landing' of the uneasy decision of the leap. The leap is not absolute abandonment, yet the 'decision' itself is a moment of rupture, an awareness that shatters the usual anticipatory frameworks. It is to be prepared for, but ultimately is a leap away from certainty, calculative reasoning and everyday representation. It is a radical acceptance of vulnerability. Therefore, as noted, here we can only proceed in a hesitating uncertain manner. Being near to the withdrawing/emerging nature of the open void beyond leaves one vulnerable. Moreover, the very two-folded character of 'leaping' (i.e. parting and joining) denotes an element of being fractured and divided. Being a turbulent region, the threshold may be frequented but itself is not "inhabitable", that is, one cannot *languish* in it, but can only 'come to pass' there[113]. It is transitory and yet it endures in its grounding significance. The momentary nature of the event of significance (*Ereignis*), with its awe and wonder, *also* confirms that we cannot abide there; its experience cannot be a continuous feature of everyday life. Yet, as a *way of being* it is a place of transition and of transformation, whereby we become, what Heidegger calls, '*more experienced*' as earthly dwellers[114].

At times we need to 'leap', yet the return to the familiar, the everyday, is inevitable. While we cannot linger in the crossing of the threshold, once its significance of originary astonishment is experienced things can never be quite the same. Truth, now safeguarded as a new kind of knowledge, nurtures the newly-found experience and imparts the wisdom necessary to make appropriate everyday decisions. To be 'more experienced' means one is *beginning to dwell*, Heidegger one is *becoming at home*. Becoming at home, in a journey that crosses the between, is a return to the intimacy of the origin. The origin is never far from the everyday. In essence, an ethos of "dwelling" means locating the unfamiliar within the familiar[115], whereby then, what was familiar for so long, may be revealed as estranging and confining; as no longer adequate for the being that is human. By returning to the familiar in a more experienced manner humans 'dwell' and 'build' in a soil now properly envisaged and cultivated.

In a largely unacknowledged manner, human beings yearn for the open of the threshold. There is a 'treasure' to be found here. However, it is not so objectively available; as it reveals, at the same time it seems to withdraw.

---

113. SE pp.152-3.

114. EHP pp.l39,42,161.

115. TAH p.32.

It is a treasure that is sheltered within a kind of tension; in the 'belonging together' of opposites, as the "inexpressible accord of nothing and Being"[116]. Therefore, instead of the vague sense of being in 'harmony with the cosmos', there persists a tension, a 'rift' that divides, closing one way of being and opening another. The unique relationship itself is explored, affirmed and celebrated in the threshold. Then we are able to return to beings, not in an empty tranquilised harmony with nature, but in the now *grounded struggle* of creative decisions, actions and works in a way of being that shelters the ground of human dwelling and the earthliness of the natural world. To be sure, the decision of the leap into the uncertainty and turbulence of the threshold risks the consolations of the habitual and familiar. However, in so doing and sojourning close to the unpresentable, the self is redeemed, beckoning the return of the gods.

The journey from the familiar to a deeper realm is not preformatted. Therefore, some thresholds may be modest small steps; fleeting unexpected moments of insight and significance in the life of the self that interrupt and disturb daily preoccupations and leave something of consequence. Thinking always has the possibility of a transformation that springs from a juncture in each individual's uncertain path when illuminated by the truth of Being. So, although the call of Being is always the same summons that grounds human essence, *not all thresholds are the same*[117]. They differ according to the time/space of the moment and the unique individual who seeks Being's resonance within it. Disclosures can be of a different order. Every situation and context shapes its own response. Although in the crossing we attempt to stay near to the manner of thinking at the limit, we are not shackled to binding prescriptions as to how it is to be done and what to expect. I maintain that there are *many* possibilities in which the truth of Being unfolds, which may be recognised in the threshold. Such moments are *traces* of the experience of Being, which may be perceived as instances of exile, a response to a 'call', a rupture with the familiar, where the realm of everyday equivalents is no longer indispensable.

Yet, even modest moments of disruption point into the direction of a *higher threshold*. This touches upon the ground of Being itself; a void with which we need to come to terms for a homecoming that is sustainable in life's journey. This is the *abyss*, the 'pure open realm' whose ineffable depth I explore in the unification of "the Grounding". Although all the unifications are to be grasped in their essential unity, the Grounding, together with the Last God and the Ones to Come, nevertheless are more directly concerned with the kernel of this homecoming. The 'Grounding' is what prepares the site for a home; the 'Last God' is to do with the recognition of the hunger of the impoverished human soul, of what is lost in the withdrawal of the experience of Being. It takes hold of the last forgotten vestige of the 'spiritual' as a deep need. The 'Ones to Come' alludes to the human

---

116.SE p.135.

117.SE p.153.

response to the call of Being. It reflects on those who are called to be the wanderers and take the modest yet essential pathway at this moment in history.

Let us dare to wander the edge of the void.

Part 3

Homecoming

Chapter IV
The Grounding - Wandering the Edge of the Void

We have moved closer to a way of thinking that could be conceived as a spirituality that is appropriate and necessary for the modern age. Perhaps, as a response to 'the call', 'the leap' has emerged as a kind of a decision; after all, the void holds both the threat of nihilism, *and* a glimpse of an inexhaustible source of deep significance. Perhaps, to affirm the latter is merely a preference. There seems no conclusive answer for or against either response to the void. This is difficult terrain fraught with philosophical traps for the unwary. It may be argued, quite persuasively[118], that intellectual and moral nihilism can be countered in purely human terms. Human beings are innately purposeful beings; they do not need to be given aims and goals from 'outside', i.e. by some supernatural interference. Although these goals are always limited, fallible, conflicting and diverse, they nevertheless result in a semblance of effective cooperation and mutual

---

118.See J.L. Mackie in *The Portable Atheist* (PA), p.253, as well as other essays in this volume.

trust in the midst of competition. Conscious purposiveness provides us with the ongoing task to devise norms, principles and values that should create a better future; a society in which peace, freedom and justice could flourish. Yet, I suggest that we are still left with a kind of existential emptiness, which has its origin in a tenuous nihilism of the soul. The spiritual

hunger I wrote of earlier persists in the psyche of human beings in a way that is not merely 'psychological'. While accepting that the universe was not designed for human benefit, the human longing to look yet further for ground, or a. This fundamental trait itself that characterises the human being cannot be excluded from any meaningful exploration of a grounding source that underpins life.

Others who have tried this inevitably run into serious difficulties. In the theological arena Hans Küng is probably amongst the most adventurous. In his vast work *Does God Exist?* he participates in this quest by an impressive and wide-ranging display of scholarly practice. However ultimately he fails in his attempt to postulate this ground via a God who somehow supplies a ground for reality. Without going into all the details of the problems associated with his hypothesising, I believe that at its core is his unwillingness to let go of the desire that a God is necessary to provide this ground. If only he would be content to contribute thoughtful and insightful reflections about the problem of grounding without trying to *prove* something that we can still call God, he would have succeeded.

The poetic imagination, which I hold to be the proper rejoinder, is neither logical nor illogical. I cannot claim that the truth it seeks expresses 'reality' any better than the truth of pure reason. It does not try to explain why there is a world. I cannot 'prove' that the image of the threshold with its abyssal ground conveys something of reality via a process of critical rationalism. Yet, it can be attested to in human life-experience, which is able to go beyond pure logic and attend to 'something more' via the innate creativity and poetic imagination that imparts value in existence. Our grasp of reality is notoriously flawed, yet that does not mean there is no reality. The human experience does not operate in utter emptiness; the void I refer to is not nihilistic nothingness. Undeniably, on this uncertain pathway we can easily succumb to reassuring but illusory conclusions. Therefore, we must always be thoughtful and alert on this journey. We need to both knowledgeable, *and* open to the other truth. And we need to depart and wander.

But what is it that we depart and where do we wander? In this venture the wanderer does not know *where* he or she is going in terms of a place and time that can be pinned down. Yet we can say that being on the edge of the void is a realm that is *foundational*. When venturing near the existential void, starkly exposed by the relinquishment of religion and the instrumentality of modern life, we wander on the border of a site that proclaims an 'event': an occurrence of existential significance. This is the pathway or the crossing of the threshold, which grounds the human essence and prepares the way for a more holistic existence. This is a journey that cannot be completed and consequently discarded. If that were possible, nihilistic boredom and indifference would soon again take hold, leaving more voids to be filled up with distractions that ultimately prove to be vacuous themselves. Instead, in this pathway modern humans approach the unbound and unlimited, the realm of 'gods' that imparts consequence in life. Its experience is thereby more *enduring* in its worth than the fleeting rewards of novelties, packaged entertainment or the proclamations of prophets and gurus.

In order to be able to continue our venture (which implicates the question of how modern humans are to meet the challenge of nihilism) we now turn to the abyssal ground itself.

The third feature of the "New Philosophical Logic" (introduced on p.54) alludes to this ground:

> This ground, that inexhaustible source for thought, is unlike that claimed by religion in the form of an ultimate being, as it cannot be possessed, worshipped, defined or represented. It also always leaves thinking incomplete. It is not an ultimate state of unambiguous knowledge. Yet, beyond its incompleteness is the truth of Being, which is unlimited and 'throws light' on the in-between region of the threshold.

> Once the imagination has exceeded the everyday limit, thinking near the highest threshold becomes 'translucent', i.e. open (although in a partial way) to the inexhaustible truth that has its origin in the abyssal ground of Being.

Here I extract those aspects of Heidegger's unification of "The Grounding" *(Die Gründung)*[119] that illuminate the place of the abyssal ground in the pathway of threshold thinking; a venture towards 'that' which grounds earthly dwelling. We shall see that the 'that', unlike a God, is not 'something' whose existence must be proven. Yet it offers an understanding that holds out an intellectual and a spiritual foundation for the journey of homecoming to one's most essential place.

At the conclusion of the last chapter I suggested that all small moments of significance allude to a 'higher threshold'. Michel Haar writes that the *highest threshold* is: "where one loses one's footing and sets foot in the elemental"[120]. This is the fathomless "ground" of the "*abyss*", beyond the threshold. As the many failed attempts perhaps demonstrate, there cannot be a definition of this seemingly dichotomous ground. Indeed Heidegger often stresses the need to be in the *nearness* of this originary realm, and for it not to be 'explained to death'. The need for a nihilistic modernity is to recover this nearness, or perhaps more urgently, the *possibility* for this nearness, to an 'openness' where the passing of gods takes place. *To be near* is the role of the threshold.

Embarking on the journey through this uncanny domain is to be done in a manner that respects its limits and recognises what is possible and what is not. It is a '*way*' of being' that honours the transformational pathway itself.

---

119. For a concise scholarly and faithful explication of Heidegger's *Die Gründung in Beiträge,* I suggest Vallega-Neu's *Introduction* (CPI).

120. SE p.154.

Here the "highest threshold" is sensed as the other boundary of what we might call the 'far side' of the threshold, near to the ground. We can neither see, nor understand or explain this ground. As Heidegger affirms, "We never *find* the ground in the abyss, because Being is never *a being*". Nevertheless, while there is no access to this ground as such, it can, and must, still be a matter for reflection. Therefore, here I attempt to recognise more clearly this source of withdrawal and concealment, the origin that establishes limits; this pure ground that in its stillness 'calls' and 'grounds' human beings and beckons as a 'hint' for a meaningful life. The pathway involves a mode of performative thinking that ventures, and is itself shaped by, the ground and its limits. Here we are "seeking the boundary's mystery", hidden in its silence[121]. For this we need to reach deeply into the pure open site of the abyss[122]. In venturing 'beyond' the limit of the everyday near the existential void we strive to 'touch' the limit of things. This undertaking does not seek for the journey to come to an impossible conclusion. To seek is not to 'solve'. Clearly, this is not a ground on which the dogmas and assurances of religion can be erected. However, neither is it a ground from which 'anything goes'. As a foundational ground it is more intelligible for the wanderer who seeks an ongoing correspondence with its awesome and wondrous nature, without the expectation of divine revelation. The wanderer anticipates the unfolding of veiled meaning and aims to remain *underway*; grounding takes time and effort.

We have seen that the pathway to be followed crosses over a threshold into a between, an unfamiliar domain that opens-up when the habitual, the purely objective and measurable, is departed. It opens up because of what lies beyond it. It opens up as it approaches the void, the abyss that falls away. This is the strange ground that presses upon humanity a kind of decision. One pathway may lead to a nihilism where the void is hollowed out as a bottomless chasm that can no longer function as a ground for human dwelling. The devastated ground then leaves the human psyche in disarray. Confused, it seeks desperate ways to satiate the void, now perceived as in need of filling. The devastation of the planet is the progeny of the laying waste of this ground.

Yet, for understanding to be foundational it needs to venture the *edge of the void*. It is on this perilous edge that *pure Being* may be sensed. To be sure, here in the passage of the in-between space that leads from the objective, material world to the proximity of the abyss there is darkness as well as light. In this strange, unhomely and enigmatic 'half-light' the imagination exceeds itself. In its radiance the melancholy springing from the inevitability of death is fortified by joy at the inexorable-ness of a life that arises from the dark earth to infinity in the fullness of possibilities. The greatest threshold, the event of deepest significance, takes place in the 'heart of the night' near the abyss. This site, glimpsed in a moment of

---

121. Cf. OWL p.41.

122. PLT p.94.

darkness, is essential for grounding earthly dwelling. Like a holiday, it needs to arise from time to time. Therefore, we need to depart and wander. We not only depart something familiar and habitual, but also leave the sub-jective and objective selves deemed sufficient for the requirements of an in-strumental world. These are momentarily relinquished so that our inner-most selves may be transformed and anchored in the void. I return to this thoughtful movement beyond subjectivity in the next chapter, but for now, as we approach its edge, our concern remains with the 'pure place' of the void itself.

This pure place beyond the threshold appears like, what Heidegger calls, an "*Abgrund*", an abyss of apparent nothingness. Yet, this nothingness is not simply 'nothing'; instead the abyss beyond the threshold is a ground that seems to withdraw as ground. The unlimited void falls-away as an in-finite originary ground of '*no-thingness*'. This abyssal ground out of which the event of consequence (as *Ereignis*) and indeed all materialisation arises, characterises most profoundly Heidegger's attempt at a thinking of "Being" that is without a descriptive and metaphysical ground. It is a ground that is 'groundless'. It is bottomless and fathomless; a falling away of ground. Yet, enigmatically, it is the ground of all grounds, whereby it grounds what belongs most essentially to truth. Heidegger's *Abgrund* is usu-ally translated as abyss[123]. I consider *abyss* as something like a metaphor for one of the characteristics of this ultimate ground. Therefore I sometimes use abyss when wishing to emphasise the characteristics of void, openness, withdrawal and vulnerability. At other times I might use different descrip-tions depending on the particular emphasis or context, such as simply the void, or the 'pure open', the unlimited, or the ground of Being.

The *ab-* prefix of Heidegger's *Abgrund* indicates a 'staying-away' of what is essential and originary to truth. It conveys a sense of a ground that stays away and "prevails as a hesitating refusal of ground"[124]. The latter means that, as soon as we *think* we have ground, it seems to 'fall away'; there are moments where ground appears to be ungranted. Yet it endures the epochs of human history as a self-sheltering, concealing and a hesitant *offering* of ground. In *Beiträge*, Heidegger alludes to differences in emphasis that may be attributed to *Abgrund*. In pronouncements worthy of much solemn at-tention by philosophers, Heidegger writes, "Der *Ab*-grund ist Ab-*grund*" and "truth as ground grounds originarily as *Abgrund*"[125]. All this emphasises the unified yet two-fold nature of this ground. At times it appears to ground the truth of existence as such, then again it withdraws.

So, Heidegger's *Abgrund* clearly is not absolute. In this uncertain ground Being draws us along by its very withdrawal into enigmatic duality and yet unity of seemingly differing aspects, such as time and space, revealing and concealing. Concealment itself is not something negative to be avoided, as

---

123. See Daniela Vallega-Neu in CCP p.79 n.25 and CPI p.88ff for brief discussions.

124. M p.xxi.

125. CP p.xxx; p.267.

it *shelters* what has withdrawn in its inherent silence. We have seen that disclosure does not merely stand in opposition to concealing. As Heidegger suggests in *Parmenides*, the *dis*closure at the highest threshold is at the same time an *en*-closure. Just like *dis*semination is not opposed to the seed and in-flaming does not eliminate the flame, dis-closure is equally for the sake of an en-closure as a sheltering of what is unveiled. This two-fold expresses an intrinsic unity whereby each partner is brought into its essence. On the one hand, it is *dis*closure as the taking away of something that withdraws and conceals, and thereby transforms and displaces an existing situation. On the other hand, it is dis*closure* as a sheltering en-closure. Therefore, in the threshold a different way of thinking pays careful attention to both the emergence of beings and to their withdrawal; what is being revealed *presupposes* the hidden nature that belongs to the stillness of the abyss. What is yet to be thought and named tends to withdraw and appear remote. That which withdraws itself cannot be *determined* as such, but we can still be near.

The other apparent duality encountered here is space and time. Space and time are normally presented as quite different concepts. However, we also know that the *spacetime* of relativity has reunited space and time in a single entity with a four dimensional geometry. Its elements are now measured interchangeably in distance and span of time. Time becomes more space-like; space more time-like. Perhaps its astonishing unity may already be sensed in the physics. Yet, Heidegger is still not impressed with its continued calculative objectification. He conceives the *Abgrund* itself in terms of time and space together and equally in a mutual 'play'. This he simply calls 'time-space', which interrupts linear, sequential time, and our everyday sense of space. Heidegger argues that *time-space* has been broken up into time and space, even though, as he conceives it, they belong together. The crude universalised measurements of 'clock time' do not represent time itself any more than kilometres and hectares represent the mountains, valleys and rivers of the landscape. Already in his earlier work he argues that clock time makes human beings lose their own time, unable to "stay" with the authentic moment, whereby we are then "everywhere and [yet] nowhere"[126]. We are unable to stay with the existentially meaningful *moment*. Determinations and activities then do not have the same potential (and 'sense of timing'!) as those taken by one who "knows his own time", that is, does not fall unawares into common time, and whose conception of *time* leaves him free to *be there, alongside* and *part of things* and to see things anew[127].

So *time-space* is not to be understood in the usual techno-scientific representations of space and time, which have become "framing representations"[128], i.e. they are only permitted to present themselves in an

126. BT 318ff.

127. Kisiel, Theodore. *The Genesis of Heidegger's Being &Time,* Berkeley: University of California Press, 1995, p.345.

objectified manner. Although these are practical for the prevailing thought patterns of instrumental activity, they are inadequate for the elemental thinking and the creative response that accords with the crossing and sojourning the threshold. Consistent with the nature of the threshold, *timespace* then has nothing to do with an actual 'location' in time and space, but instead emerges out of a pre-objective region of spatiality and temporality. Here we attempt to grasp space and time in their pre-mathematical form, which means these are envisaged more on the basis of an *event* of 'spacing', which provides or 'frees up' an opening for truth to unfold. The difference between space and time is not effaced, but is in fact enabled. Here, time becomes essentially dispersive; it is what provides 'space', with space having a gathering/withholding character[129]. When the imagination overflows in the threshold, the pre-objective unity of time and space is sensed as an offering to which human receptivity and creativity may respond[130]. Time and space retrieve their primordial accord in this 'in-between' realm. Here they are 'released' and allowed to interplay in a kind of poetic presentation that may be witnessed by those whose imagination allows a kind of beholding.

Transcending the limits in the threshold is then not towards some timeless realm, or even the timelessness of 'a permanent now' seized in an illusory state of pure consciousness, as ostensibly proclaimed by the gurus of "Now". It is rather a step back into the opening of time itself. In the strange dimensionality here, pre-objective time and space in a sense 'rest in stillness', in the *Abgrund* or in the 'not' of Being. There they await the encounter of Being and beings. In the experience of the threshold, momentarily freed from the relentless time of representation, human beings sense the unity of the existential *relation* between the phenomenal and the noumenal as Being itself. Taking up George J. Seidel's deliberations[131] and adopting a somewhat speculative course, I suggest that in the separation between things and Being there is a gap, a "noumenal nothing" through and beyond which the joining of truth and Being occurs. It is where 'Being happens', so to speak. Perhaps, at this juncture 'the truth of Being' reaches out from beyond the phenomenal, that is, from the abyssal ground. Within thinking it remains possible to traverse this noumenal gap, or 'overflow' a limit to a 'region', a 'threshold' where this unification takes place. Here the meeting and union of truth and Being escape the reaches of everyday space and time. Because human beings are uniquely capable of this kind of thought there resides within us the possibility to grasp such a moment of truth that is beyond the purely sequential model of time and quantifiable

---

128. CP p.260.

129. TWH p.199 n.21.

130. Cf. TWH pp.143ff.

131. See George J. Seidel's discussions in BNG pp.43-87.

space. Within the "inner life of self-reflective reflection", human beings can thereby think *the unity itself* that rests within the truth of Being.

In this meeting that takes place in thought, the human objective and instrumental inclination is relinquished in favour of poetic imagination. This event neither occurs *in* time nor in eternity. It comes into being by the possibility and the very need of appearance and change. Conceivably it is this *need and possibility* residing in the *stillness* of the abyss that is then offered up to thought, as the relationship between Being and beings moves in the direction of the unveiling of truth. The converging lines of the direction of truth and Being that meet in *time-space* is a realm which only human beings can join; it is only 'there' that this event can take place. In thought the encounter with truth happens. This encounter contains elements of both coming together and moving back. It thereby holds the origin of the strife and turmoil of the threshold. In this encounter, as Being and human being move towards one another, 'something' overflows the abyssal ground. This 'something' is the event of consequence (*Ereignis*) in which the bridge between Being and beings is erected, allowing these to emerge in their essence: as 'gods', humans and earthly things.

There we are open to what is essentially silent and hidden: the self-concealing ground and self-sheltering truth of Being. In this transformative experience beings are perceived as an enactment of Being itself, where it throws its light unto what is happening in the threshold. Beings, limits and boundaries are concealed and unconcealed in this single event that holds both unity and movement. In the movement away from representation, logic and everyday instrumentality the imagination is enabled to grasp something of the pure openness of *Abgrund*. Yet it retains an adequate hold of objectivity and rationality to respond appropriately and creatively. In this encounter it displaces and yet generates the presentation that makes the everyday possible. In the suspension of familiar objective spatiality and linear temporality, the essential nature of things may be brought out more clearly in the creative response. Beings as such may emerge as a distinctive unity of present, past and future; a distinctiveness that, due to a kind of lack of interest in the essence of things, is usually denied, whereby these can only withdraw into ordinariness.

Because of its unfamiliar abyssal nature, it is not surprising that, in the instrumental attitude, the ground that grounds (*Abgrund*) remains forgotten and unquestioned. It is the buried under, what Heidegger calls, the *unground*, the 'unessential' or lack of ground, which negates all sense of ground, whereby the truth of Being remains hidden and unexperienced. Whereas *Abgrund* is an *affirmation* of ground, *un-ground* is a *no* to every ground. *Un-ground* becomes apparent in a human comportment which is only able to think in an instrumental, objective and utilitarian way. There is then no sense of *care* about the deeper essence of things. The un-ground disassembles or covers up *Abgrund*[132]. The emergence and taking hold of

---

132.CP §188.

this un-ground, and the consequent distancing from unveiling emerging Nature (as *physis*), has resulted in reality being seen only in terms of object-ness and objectivity[133]. It shows itself in the way technological thinking becomes overpowering and all beings are regarded as mere resources. Even the *realisation* that the loss of Being has taken place, and has its costs, is not possible in the scramble to evaluate everything objectively, to become ever more efficient and to worship at the altar of growth[134], fuelling the consumption that attempts to fill the vacuum the un-ground leaves in its wake. Therefore, un-ground therefore cannot be interpreted as 'nothing' either, but needs to be recognised, revealed and maintained in its proper place.

The encounter with the nothingness of the abyss touches the infinite, creative source (not the explanatory cause!) of all that is; it is one with pure Being. Using the term "infinite" here implies that thinking in the threshold indeed senses the 'open space' beyond as *unbounded*; an infinity in radical excess of any calculation. The infinity of calculative thinking is *not* foundational, because, as the *void* cannot be discerned on its own terms, it is excluded from such thinking or it is merely abridged to o or ∞. It is thereby "levelled off in advance unto the same of what is countable and what makes counting possible", asserts Heidegger[135]. Yet the void itself is indifferent to calculative demands for control; it endures as the un-representable silence from which humans cannot simply escape. Here, the infinite is not as an ongoing progression or a collected whole, that remains endlessly subdividable. It is a un-limitation that is not calculative, but involves a kind of gesture or offering of the infinite itself. It is not the "infinite sprawl" of absence, endlessness or eternity, but rather becomes the infinity of a boundless beginning[136]. This, I believe, encapsulates what Heidegger calls, the '*other beginning*'. Here the possibility of unique encounters with the gods endures. In the threshold we gain access to this union of the unlimited with thinking itself. For Heidegger, this is what really matters; not so much the difference between Being and beings (referred to as the ontological difference), but the *relation* itself, the *belonging* of Being and thought, arising out of the void. It is this *relationship* between "man" as the being capable of reflection, and Being, rather than the components of the association that is at issue here[137]. This then is how human beings *ground* themselves. Not by finding '*a*' ground as such, as claimed by the proponents of an assortment of

---

133. CP §61ff.

134. Heidegger discerns three concealments that participate in the abandonment of Being: "calculation, acceleration and massiveness" (CP p.84), which results in the diminishing of the self-understanding of human beings in an authentic relation with being.

135. CP §241.

136. FT p.223.

137. ID pp.7-8.

religions, but by grasping the connection itself, as the essential belonging together or the unification of the boundless with the imagination.

In the threshold near the void we 'throw the dice' upon the pure open site. Here the joining together of the intangible event to the vulnerable circumstances of the crossing is accomplished. On the brink of vanishing familiarity we encounter the decision whether to flee the mystery with its 'unhomeliness', or to 'go under' into the abyssal site of the void. This is the 'sublime experience' (to which I return shortly) that is offered at the limit of human imagination and flows onto the between. In the threshold we move 'against the grain', towards that source that is the open expanse from where all things reveal themselves. 'The Leap' I wrote about in the previous chapter, is the decision to "throw the dice" upon "the oceanic site"[138]. It is not one of choice; that would merely be 'taking a chance'. I am not talking 'numbers' here. It rather opens up *possibilities*. As Mallarmé wrote, "every thought emits a cast of dice". Therefore the 'decision of the leap' is from a standpoint that is itself undecidable. Throwing the dice also 'catches' something, namely the gesture of the infinite; the gift of Being itself as it unveils the truth that is hidden within it. To shun the gesture of the infinite would abolish its site; thinking would be unable to overflow the border of the everyday, remaining shackled to the restraints of the seductive powers of calculative and technological thinking. No event of consequence will have taken place. Yet, it does not depend on us whether the event takes place; it occurs on its own. However, we can *decide* to 'watch over' it, meditate on it, delight in it and express it in diverse, uniquely human ways. This is the strangely inexorable, 'belonging-together' relationship of the essence of human beings with the abyssal void of Being.

Here, at the limit of our wandering we are in danger of being overwhelmed by the apparent nothingness of this realm. Bordering on the edge of the void leaves us vulnerable; we are only just 'hanging in' near the stillness and unfathomableness of the ground that falls away, like an abyss. It leaves one, in mortal vulnerability, 'holding out' the self towards its strange dimension, perilously unprotected and unguarded. No longer can we fall back on the reassurance of objective knowledge and familiar concepts that can be represented in a straightforward manner. Sojourning on 'the edge', in the nearness to 'the pure open', *the very limit* of the highest threshold itself is truly encountered as it touches and transforms. This feeling is not a familiar emotion of the self, but becomes more an experience of an awareness that one is undergoing something *foundational* that is beyond the limit of oneself. The experience of the threshold is the contemplation of "sublime presentation" at the limit. This meditation, that Jean-Luc Nancy names the "thought of the sublime", cannot enter the (pure) space beyond the limit, which is 'the Whole' or the ground itself[139]. However, in the

---

138.Cf. Alain Badiou's Mediation on Mallarmé in *Being and Event,* New York: Continuum, 2005, pp.191ff.

139.FT pp.235-6.

threshold, it 'comes to pass' when the sublime of Being 'overflows' from the void. There is nothing beyond the limit of the threshold that is present-able, whether positively or negatively, except for the *"thought of the sub-lime"* itself, which is 'glimpsed' from the abyss, the retreat of ground. Here one seems 'stretched' to the limit. The exposure to, and appropriation (one is affected and transformed) by, what is beyond this limit, i.e. the open abyssal nature of the *Abgrund* itself, marks out the proper dwelling place of the human being. The no-thingness of the abyss, which holds emergence in reserve and in its care, is taken as the starting-point of all human characterisation.

Human beings, in fidelity to the relationship with the void, are then the momentary site, the 'between' for the grounding of the truth of Being. It is *momentary* because its ground always features the retreat of ground whose excess 'overflows' in the *Ereignis* event; a moment of transformative signi-ficance, where the extraordinariness of all beings and of life itself is sensed. Discerning the otherness or numinous of the open void is a 'feeling' that is without concept or affect, but grasps something in what is sometimes named the sublime experience. In the next chapter I shall have a closer look at the nature of this experience, where we undergo the expansion and liberation of the 'soul': that extraordinary, astonishing human characterist-ic that is attracted to and concerned with being and is able to reflect on this. The threshold is the domain of *wonder*, as a fundamental disposition that is open for the truth of Being. "The Grounding", enables the experi-ence of the strangeness and originality of the human relationship with Be-ing. It sets us free as the unique beings that the resonance of the echo calls us to be. The Leap itself then has become ground; it grounds the *self* of the one who knows that his or her individual dignity is affirmed in this re-sponse. Such authentic human being is needed for the occurrence of *Ereignis*; we are its site, the place where it can 'take place' as the meeting point of its meaning and relevance.

To approach the 'reserve' that is held in trust in the concealment of the abyss, implicates the 'something more' needed in the ongoing human hun-ger for meaning. It requires the 'risk' of venturing outside familiar territory; hence it is not for the fainthearted. However, this cannot occur when hu-man beings place themselves at the centre of all activity; when it seems that absolute knowledge which 'explains' everything about us is all that matters. The inner preparedness that all human beings retain for the encounter of significance then remains hidden, culminating in a groundlessness where we believe the ground from which we speak to be borne solely from our own lived experiences. Yet, when released from the hold of the ego, the reticence or modesty of this attentive readiness allows "what is secret to be secret; the infinite loving tenderness that *knows how* to let the enigma of the most familiar things appear"[140]. It retrieves a sense that life and its earthly sanctuary are deeply meaningful and deserving of faithful reverence.

---

140. SE p.146.

In this section we have approached the uncertain ground of our being and the manner of its revealing by a "logic-in-the-making", which seems more primordial, more poetic and meditative than everyday rationality. Yet it does not abandon the latter; here different realms meet, enabling an exchange of truth to take place in the overflowing of limits. We have seen that this logic is self-opening as it is inherently translucent, or open in an osmotic kind of a way, to that which lies beyond its own incompleteness: the truth of Being, itself arising out of the inexhaustibility of the abyss. Heidegger suggested that the 'gods' rest in the stillness of this void. As already noted, these are not Gods resting on heavenly thrones, dispensing judgement, commanding worship and 'causing' things to happen. It is time to call on these strange gods.

The Gods that Save

When that cherished poet/philosopher/artist Michael Leunig is asked the "bizarre and bewildering question" of whether he believes in God, his response is to "go peacefully blank"[141]. He recalls that as a child God was simply a useful word that expressed something vague and mysterious, yet significant, like life itself. He was quite content to innocently accept this word until "out there, in the culture, he was confronted with the difficult idea that God had created everything." God became a being, made in Man's image; a figure of absolute authority who demands our deference and who we should approach with solemnity, earnestness and trepidation. This loss of innocence, enforced by the vanity of the current age, mirrors the loss of primordial awe and wonder at the mystery of life that the Greeks experienced, as representational metaphysics gradually took its place. Of course, for Leunig, as it is for many, this is not a conception of God that really makes any sense, nor does it solve any mysteries. For him 'God' cannot be grasped scientifically, rationally or theologically. Instead he refers to "a sense of God that can only be held lightly and poetically... this sense is the natural and delightful fluidity of spirit we call poetic imagination". Leunig is deeply aware of a sense of the numinous that does not attempt to explain, define and represent, but is to be approached openly (blankly), so that the imagination can exceed the everyday. Authentic creativity is involuntary, yet it involves a human essence that needs to be 'released' in order to encounter it. Leunig understands this vital essence as our "innocence". What is this innocence? Does it involve child-like naivety unspoiled by the ways of the world? Childlike yes, in a sense of being open to what is being revealed without presumptions. Yet at the same time I believe that it is good to be knowledgeable, so that we can remain aware of what is happening here. The innocence we seek involves a return to roots; a realm where one remembers primal poetic instincts. When the mind is lured into a way of thinking that is wholly restrained to rational objectivity,

---

141.M. Leunig, op.cit., 'God only knows', p.233.

it loses this innocence. Leunig's writings and art reveal that he truly is a frequent venturer of thresholds.

Heidegger thinking of "Being" also attempts to overcome metaphysical representation, but, can we say that Being ultimately is just another name for God, albeit now a name for what is non-representable and indefinable? It certainly would not be a God that requires quarrels, wars, inquisitions and other troubles to settle perceived differences as to 'his' precise nature. Yet, to declare that 'Being really means God' is quite naïve and a gross simplification of Heidegger's attempts to recover for thinking, what theology has often regarded as its exclusive possession. Such questioning has been lost in the objectification and metaphysical representations of religion, or stifled in unconditional and dogmatic belief.

Whereas ideas about the divine and spirituality usually reflect anthropomorphic, objectifying thinking and metaphysical preconceptions, Heidegger, in his post-theological project, attempts to avoid and overcome these in a way that is neither theistic nor atheistic. Despite the use of words such as 'soul', 'spirit' and 'the holy' he consistently aims to keep theology and philosophy separate, so that his thought would not be a secularised theology. His approach in *Beiträge* endeavours to maintain a distance from the unconditionality of faith; indeed as what he describes elsewhere as "two spheres separated by an abyss"[142].

Heidegger does not think of Being as a being, or even as a *concept*, as it is essentially hidden, unnameable, and un-representable. Therefore he certainly does not regard a belief in an objectified metaphysical God, often grounded in fear, as philosophically adequate. In his analysis and pathway of thinking, the existence of a metaphysical entity beyond human experience is the realm of conjecture. As theology eventually was *derived* from the originary thinking that Heidegger attempts to retrieve, one could say his pathway is theologically indifferent. Religious belief is a *secondary* human activity that has its primordial roots in the reality of human finitude. God as the most transcendent (*maxime transcendens*) among beings is for him an inadequate notion and therefore is not considered. He believes that God as an objective entity cannot be questioned philosophically. Indeed, pretensions to a familiar, 'personal' God are to him irreverent. These are attempts to personify the mystery that can only wait beyond description for poetic mediation in a manner that preserves the truly holy. He regards a god that is susceptible to representation or evaluation as nothing sacred. If it is within human power to procure the transcendent, then the holy would be diminished. To be sure, for Heidegger, a moral, redemptive, metaphysically constructed God is dead. For him the thinker and the poet are nearer to divinity than the faithful believer. Only if thinking surrenders the metaphysical God does it become free for "the divine god."[143]

---

142. WCT p.177.

143. ID p.72

Often God is merely that which 'explains' anything still not understood by science. Yet, Heidegger questions the pursuit of absolute certainty commonly held by both science and theology. To treat Gods themselves as an explanation of what is, in the end, not explainable, is for him a "de-god-ding"[144]. The gods, as he perceives them, take flight in the face of such attempts at vindication. God as an answer to a scientific gap in knowledge is nothing divine per se, if the divine means the sacred, the wondrous and the awesome, and ultimately the truly unnameable. The divine being of metaphysics is named in order to dissipate the mystery. In the history of metaphysics such a representative determination of the godly has resulted in a plethora of versions of theism. Indeed, as the metaphysical Gods harden into objectification the *"flight of the gods"* is hastened. Heidegger suggests that the truly spiritual person, facing the ultimate mystery, is struck with awe and responds with a "silence" that is proper to stillness of the void[145]. He laments the "atheism" of the modern world with its "absence of gods", which refers to the utter "oblivion of Being"[146]. The gods have fled as there seems to be no one with sufficient poetic imagination to 'open up a space' for them. He therefore insists that, for the authentically "holy" to again appear and for the language of Being to be heard, the task is to remain near to the withdrawing of the gods in care-full attention[147]. This means that we need to be aware of the way that the instrumental attitude, as discussed in Chapter 2 and elaborated as a kind of 'covering over', makes it ever more difficult to sense the presence of the gods. The significance of the human relationship to the mystery of the sacred, as an abyssal stillness where the gods rest, needs to be recovered for thinking.

Heidegger's gods have their wellspring in the gods of the primordial Greek writings. In particular the "goddess Truth" of *Parmenides* is central to his conception of Nature as 'a whole', i.e. as *physis*. The loss of the clarity intrinsic to the early Greek sense of the fundamental character of beings and of emerging, un-concealing Nature has resulted in the perception of nature as merely an object of utility. Constrained by our expectations it becomes a "contrivance", a product of human making, to be managed, manipulated and used up[148]. If we instead grasp the primordial essence of Nature as *physis*, the free enduring outpouring of an elemental truth which reveals itself in a way that always keeps something back, we are beginning to understand the truth of Being itself.

---

144. M p.212.

145. Heidegger, *Towards the Definition of Philosophy*, London: The Athlone Press, 2000, p.4. These words from the younger Heidegger about silence in the face of ultimate mystery (somewhat akin to Leunig's "going peacefully blank" provide a direction for his later thinking.

146. P p.112.

147. EHP p.46.

148. BQP p.122,

Every era has sought 'the truth', resulting in many entangled viewpoints. However, here our concern is something no less inexplicable, namely the *'essence of the true'*. As discussed in Chapter 3 Heidegger believes that the deeper meaning of the fragments of ancient Greek literature is lost in conventional translations, which are dominated by the derived form of truth as correctness via the correlation between transferable situations. Thereby, for him, what is implied by *Parmenides* as the "goddess 'truth'" becomes misinterpreted as, either particular forms of irrational lived experience ('fairy tales'), or "personified" into a godly figure[149]. When we read of the goddess in *Parmenides*, we immediately transfer the image of a being onto what seems an abstract concept. However, when personified into a celestial figure she is readily disposed of in the modern world as merely a mythological god of the ancient world. How much more divine and enduring is this goddess, when understood as the very essence of truth! Our customary understanding of derived truth *belongs* to this essence. The goddess brings all things into view, placing them in the light of Being and bringing about the essential *relationship* between humans and Being.

The goddess is therefore not simply a poetic means to introduce the notion of 'truth'. Instead, Parmenides names the goddess as the essential place where the thinker as thinker is *at home*. The goddess is originary truth itself, perceived as that mystery and truth of Being which both unveils and withdraws. That is why Heidegger translates the word *aletheia* not simply as the conventional 'truth', but by the more literal 'un-concealedness', which may express something of its uniqueness if we are 'attuned' to it. As explained in Chapter 3 this word is not the outcome of a mere word-game but needs to be experienced in the 'carrying-over' of a fundamental encounter; one that transports our being into the realm of a transforming truth. As Heidegger writes, "this transportation appropriates us *into the care of the word*". Here originary language 'speaks' to those who are 'open' enough, by the 'letting-go' of instrumental objectivity. The very essence of the truth of Being is then expressed in the word *aletheia*. Heidegger considers that this emergence into the open has the "character of appearing and of shining". This light of Being is what enables us to see. The eyes can look, but Being's light lets us see. The human soul or psyche is founded on this 'seeing', and in crossing the threshold the "eye of the soul" sees beyond mere things to behold 'gods'[150].

As stressed, it is always essential to keep in mind that the "gods" of *Beiträge* do not refer to entities, or to any kind of being. Also these are not something *above* Being, but rather an *occurrence of* Being. They are 'immortal' in the sense that they embody a unity as a real feature of the world which is deeply and enduringly meaningful beyond the lives of individual mortals. When Heidegger writes that "the gods have fled", this means that a sense of the sacred, one that brings into play a spirituality that

149. P p.10.

150. P p.146.

does not invoke a being above beings, has been erased. When such think-ing is crushed we can no longer grasp the fullness of being human and the primary experience of existence itself. This kind of spirituality is not taken seriously in the modern world where discourse seems to be unable to step beyond the rational, the objective, instrumental or representational. The consequences of this are sadly obvious. Buried under all the propositional thinking and ossified by instrumental delineation and objectification, the gods' *need to essence* has become truly pressing in today's crisis of meaninglessness.

Heidegger in the thinking of *Beiträge* attempts to retrieve the gods. Here he surpasses metaphysics in a way "that turns transcendence back to the earth, remaining faithful to the earth that the gods call forth ... it at-tempts to listen to the earthly echoing of distant singing of the gods"[151]. The remote song corresponds to the 'call', the 'echo' and 'hint' of Being. To 'hear' such a silent voice requires a certain 'earthly' comportment, such as the poetic imagination or the thinker's reflections. Heidegger writes of the hint, the onset and staying-away, the arrival and the flight of the gods[152]. The flight of the gods alludes to the explicit experience of the self-with-drawal and forgetting or abandonment of Being. This is not the kind of for-getting where there is the total erasure of something from memory. Yet it may become *totalising* by way of purely technological thinking that is able to *withdraw any sense of human grounding*. This is the *un-ground* of machina-tion, introduced earlier; the denial of ground. We have seen that machina-tion is the manner in which the forgetting of Being and the instrumental attitude takes over in contemporary life. It brings about the flight of the gods. The modern malaise *arises out of this flight*. Nevertheless, in their de-parture the possibility of return remains sheltered. In a Preface to *Elucida-tions* Heidegger writes that in Hölderlin's poetry "the holy... speaks of the flight of the gods"; and that in their flight *they remain sheltered until human beings are willing and able to venture closer and dwell in the nearness of the gods*[153]. This is the hope we must have as we stand at the crossroad of the modern predicament.

The region of their flight (near the abyss) shelters the enduring possibil-ities of meaningful being. There at "the outer border of the place [the 'remote side' of the threshold] the God of gods appears"[154]. But surely Heidegger here speaks of theological divinity, especially when then insist-ing that "no human calculation and activity, in and of itself" can bring about a change in the world's woes? Is Heidegger not expressing here hu-man helplessness? No, he clearly refers to one fundamental underlying problem: "the whole of man's activity has been stamped by this world

---

151.Günter Figal in CCP p.172.

152.CP p.288.

153.EHP p.224.

154.Ibid.

condition and has come under its power". He fears that in the modern world the overwhelming by technological thinking is becoming absolute. Does this not ring true today? Climate change is to be dealt with by a bewildering array of 'fixes'; we'll do anything except transform the heart of an unsustainable way of being. When humans are no longer able to *grasp* the truth of Being, it is not possible for "the holy", the "God of gods", the wisdom grounded in the sacred so lacking in the world, to appear. The holy is needed to ground, what Heidegger calls the other beginning, i.e. another way of being that is truth-full authentic earthly dwelling. He writes that "To *grasp* means to name the high one himself", but "holy names are lacking"[155]. Here he refers to letting "the holy", the being of Nature" as totality and ultimate mystery, appear and naming it poetically (not as a metaphysical representation) in its light. Yet, today it seems that its "originary wholeness" and "awesomeness" is no longer able to confront us with its unfamiliarity[156]. The flight of the gods is in danger of becoming *absolute* through the abandonment of the ground of human beings by unbridled machination. The consumption of technology and its progeny is now perhaps fulfilling the adage that 'one is what one eats'. To be sure, technology enables us to achieve much that is of benefit, yet what is missing today is the understanding that all human activities require the gift of what is pre-given, that which is not of our own doing. The inability to accept mystery and ambiguity, the overriding human self-importance, the fear of finitude and the indifference of boredom, the averaging out in mediocre and calculative thinking, all this contributes to the inability to encounter the gods.

What Heidegger calls *"the last god"* is the withdrawal of the experience of Being that leaves *a passing hint* (one that does not simply *appear* and then is forgotten), experienced in a kind of distress or need. This distress need articulation; not evaded such as in rampant consumerism. "Last" refers to the most fundamental self-negation. However, if this withdrawal, or the not-granting of a sense of Being, is truly experienced, the hint is not an end. As it turns and reverberates into itself, it rather initiates another beginning. It becomes an echo of the gods to which we can respond. So the withdrawal is not of some entity, but is an event that has the distinguishing trait of withdrawal or abandonment, because its ground is like an abyss. As it hints of the essence of the uncertain, yet authentic, ground of human beings, this retreating is not something negative to be avoided in a quest to remain on familiar and habitual territory. The hint of the last god both conceals and shines. It grounds Being itself, but needs the sheltering by the human being, or else it is obliterated, lost to machination. This can only occur *through* us, i.e. it is *not* a transcendent occurrence beyond beings, but rather takes place within an open 'space' cleared by those who are to ensure the preservation of the 'hint' that is essential for the possibility of the arrival of the gods.

---

155. Ibid, p.45.

156. Ibid, p.85.

Like a glow out of the darkness of the abyss, "the last god" is then the most remote call; it is now the highest distress as it signals the last hint or trace of the truth of Being that can break into the stranglehold of techno-logically objectified 'man'. The technological and instrumental attitude is essentially one of 'violence'; the imposition of the will upon nature. In the abstract reign of modern technology the natural can no longer appear out of its essence[157]. The arrogance of futile attempts at placing (a privileged se-lection of) human beings alongside God (imagined or otherwise) may well be the final move towards the utter annihilation of the human essence, if not the human race. Many of the beings with whom we share the earth have already fallen victim to the greed that grows out of a dislocated sense of self-importance. When *all divinities are eliminated*, the tendency is to grasp absolute assurance for oneself, to think *oneself* as the divine being in control of all things. Then to become the master and owner of nature can only mean *overcoming* it, as in this context it now represents the ultimate lingering mark of mortality. The dangerous illusion of conquering uncer-tainty and risks may be encountered today in the full range of modern dis-course; from the cranked-up hyper-active cheerfulness of the FM radio presenter, through the predictable managerial language of corporations, to the empty rhetoric of conviction and reassurance of governmental spokes-persons. Everywhere we witness the struggle for power; yet it is it is power itself that is determinative[158]. It dominates all those who want it so desper-ately. This struggle does not understand that it is in the service of power.

In the "utter relinquishment of the ontological residues of the gods" the last vestige of the originary experience of Being is erased[159]. The self-loss of one's essence takes place as the *exhaustion* of what is truly human. The gifts of 'awe and wonder' associated with the gods are lost in their flight, the consequences of which the manipulative degradation (the "exhaustion of beings")[160] of the earth is a clear symptom. "The last god" then is that ulti-mate forgotten trace of the 'spiritual'; that "greatest need" for originary ex-perience that retrieves the fundamental human freedom and must be al-lowed to 'play' in our thinking if we are to be 'saved'. The last god is only an *end* when human beings have torn themselves away irrevocably from the truth of Being as such. If recognised as a crucial issue, it may be the first step to the other beginning that attends to the root of modern social and environmental problems. When one knows that one has forgotten the most hidden need to seek a human ground, the prevailing *lack of need* is re-cognised and reversed. For this we need to hear the "echo of Being", where the hint of the stillness of the gods passing is experienced in their uncanny

---

157. For an analysis of Heidegger and the 'framework' of technology see R. Rojcewicz, op.cit.

158. EP p.102.

159. David Crownfield in CCP p.218.

160. CP p.293.

manner of appearing and withdrawing. They "assign to man a stiller [less self-important and restless] nature"[161].

The gods 'rest in stillness' in the abyss until their echo is recognised and the essential 'belonging together' of humans and gods is retrieved. The dormant gods await human mediation in the silence of the abyss. By opening up our minds and senses, human beings 'make room' in the threshold so that the godhood of gods can happen in a kind of "mutual beholding" or in a most intimate self-belonging[162]. Gods, now emptied of substance, are effaced in this un-representable realm. They are then no longer the disfigured images of ourselves, rarefied and glorified to become objects for worship and consolation. No more are they the feeble explanation as the 'cause' for being. In the contemporary illusion they are only reckonable out of objective beings, by whom they are 'possessed' and thereby grow into super-humans; therefore "no god appears any longer"[163]. No, here, they are retrieved from a barren situation whereby they had become 'impossible'. Divinity is salvaged when the void opens up the truth of the mystery of Being. It cannot be grasped, yet may be touched when we let the gesture of the sublime offering come to pass. The setting free of beings and Being transpires in this encounter. In the "between" of gods and humans the "distress of gods" (their need for Being) is abated. Human beings, cut free from the hold of machination and the constant desire for manufactured experience, find their self-being in responding to the gods' need[164]. This kind of thinking moves onto new ground and rediscovers the forgotten 'remote' ground of human beings. In such care-full Earthly dwelling, transformed human beings are thus able to re-open the space of 'the between', the necessary space between Earth and world, between gods and humans, between Being and beings. Thinking the dimension of the godly affirms the need of both the gods and authentic human beings. The "need of the gods" arises out of the withdrawing, hidden, silent and abyssal nature of the ground of Being. The hint, onset, passing and flight of the gods affirm that we do not only need 'Being' as an event of deep significance, but human beings are indispensable for this event to occur and reveal the truth of Being[165]. The 'call of Being' has its origin in the need of the gods to come to pass in the only site they can be: human beings who in faithfulness provide a kind of clearing for this truth. Such individuals, by responding in an authentically human way, then become the *grounders* of truth. They are "The Ones to Come" (*Die Zu-Künftigen*), to which I return in Chapter 6, who ground the originary *time-space* for the departure and arrival of the gods.

161.OWL p.196.

162.M pp.218,225.

163.M p.224.

164.CP p.331.

165.Cf. CP p.288.

The 'death of God' has its correspondence in 'the flight of the gods'. The modern sense of emptiness in the wake of the death of God may well be the wake-up call needed to take greater heed of the enduring human yearning to seek the unbound. The unbound has its origin in the void to which human beings belong. Somehow the void tempts in its concealed calling. Yet, as Hölderlin discerned so well, there is no simple clear pathway; the way is crooked, there are dangers, and, as the site is not without limits, we do not travel unencumbered. There is no room for nostalgia; thinking here does not seek a self-made sanctuary, but allows the self to be shaped by the event of Being, where something of significance 'touches' the venturer. Here, at the 'highest threshold' what touches is *existence itself*. This intimate proximity of Being affects, awakens and moves.

In the threshold there is already a kind of 'opening up' taking place in response to the "highest threshold" near the pure open of the void. The threshold holds the gifts of the "echo" as the call of Being; it holds the "hint" of the gods. Although the hint of the gods and their 'saying' occurs there, the threshold itself is not *Abgrund*; we are in the crossing, where we find an openness that is near the "pure space" or "place", of "the Other", "the Whole", "the Open", the numinous or the "sublime". Such names adopted by thinkers and poets, all these signify to the Being of beings as a void or abyss-like ground[166]. The names given to its *experience*, such as 'pure perception', ultimately allude to the 'sublime experience' of the truth of Being as a transformative non-representable occurrence that is beyond subjective and objective understanding. The experience entails the 'poetic imagination', which becomes accessible when the everyday imagination overflows the boundary of the familiar in a movement beyond subjectivity. This movement brings into play what we have called 'the soul'; that innermost human essence that has become so impoverished by the pursuit of the obvious and the fleeting, the material and the self-promoting. I now have a closer look at the way thinking becomes less subjective in the threshold and more open to another truth, allowing the flourishing of the poetic imagination.

166.Cf. SE Ch.VIII.

Chapter V
The Heart beyond Subjectivity

When Heidegger speaks of subjectivity he does not directly refer to a characteristic belonging to human subjects through which reality is perceived according to the 'attitude' of the individual. He rather emphasises its fundamental nature of *dominance*, whereby human beings become objects *for* subjectivity. As subjectivity then establishes and fashions us, we lose our being; the very essence to which we belong. It closes off other possibilities of being and the experience of Being itself[167] . Yet, Heidegger perceives that the primordial Greek beholding of emerging, un-concealing nature did not involve the certainty of truth as correctness and the self-assurance and relativism arising from subjective opinions of a self-conscious subject. This manner of discernment has not disappeared irrevocably in the modern world. As he writes in *Parmenides* (p.19), its *resonance* persists today in the infinite richness of the possibilities of being. The domination of subjectivism may still be overcome at the limit; the horizon of understanding where the sensible fades and everyday imagination begins to fail. The 'subject' who

---

167.CPI pp.160-2.

has 'moved' to the limit is then touched by, what Jean Luc-Nancy calls, the "sublime imagination"[168]. Here we delve a little deeper into this change of our inner being that kindles the flame of poetic imagination.

When the imagination overflows into the threshold and senses the pure open space of the abyss of Being, there is a kind of opening up taking place. The between of the threshold involves a transformation, which at its core involves a conversion of the subjective self into a way of thinking that is less concerned with feelings and viewpoints moulded by culture, tradition, social conditions, etc., so that another manner of discernment becomes possible. When subjectivity and objectification loses its grip, thinking becomes more open to the essence of the true. What is the nature of this 'openness'? The open nature of the 'space' of the threshold has its origin in the immeasurable openness of the abyss which in a sense exceeds itself to encounter the more open imagination that has spilled over the boundary of the everyday. This then becomes the 'experience of the sublime'. We return to this experience of the abyssal void in a moment, but firstly let us reflect a little deeper on the openness of the threshold itself with which the imagination is deeply implicated.

'The Open' has been named a "pure place", a realm that is sought away from the constraints of subjectivity. Michel Haar suggests that this pure place "is an unpossessed and unpossessable place where every mark of ownership, and every limit is effaced; an elemental place"[169] . Effacing here means a making indistinct so that their 'overflowing' is possible, rather than the elimination of limits and spheres of possibilities. It is an originary realm; one where, as we move nearer to origins of dwelling, the abundant open space beyond is sensed. This pure place, approached in the crossing, is not some passive substrate. It would be an inert, empty, colourless and flavourless absurdity, whereby the authentic collapses into itself in dissolution[170]. Its *purity* cannot be a *representable identity*. Yet, because it is not a positivistic 'nothingness', this un-representable pure open place may be thought and sensed, although clearly not in a familiar everyday manner. Therefore, whilst recognising the limits to thinking and language, *abyss*, as the pure space of fathomless Being that 'falls away' as it is approached, seems an appropriate metaphor.

In Rainer Rilke's poetry, the *Ur-ground* is such a 'pure place'. It is "the Open of Being", which expresses the pristine ground of beings as that which grounds beings. This pure ground is the being of beings in their totality. From this ground Nature emerges in a liberating surge from its concealment[171]. Rilke's *Ur-ground* clearly has much in common with Heidegger's *Abrund*. The venture of world-disclosure, of life and nature as a whole,

---

168.FT p.233.

169.SE p.149.

170.Cf. FT p.284.

171.PLT p.99.

is released from this holistic domain. For Rilke, "the Open" is that which does not obstruct, being itself free of barriers. It 'admits' something that already attracts (Being and human being) but is kept apart by the instrumentally objective hurdles we erect in the restlessness of the everyday. So, in this open something is brought together that already belongs. We can recognise here the correspondence with the threshold, where the imagination is able to overflow the barriers accomplishing the human role as the 'open site' where Being is able to essence.

The poet's pure site is the divine regenerating and healing centre of things, as the un-nameable and un-locatable place of 'the Holy' that speaks of the gods. In *Beiträge* Heidegger does not oppose such a portrayal of this un-representable realm, when he describes *ur-ground* as the resonating site where Being 'happens' in the event he calls *Ereignis*[172]. Heidegger showed much interest in Rilke's poetry, yet he had some reservations about his "Open". He considers that Rilke overlooks the more essential 'light of Being' itself, i.e. the pure unveiling of beings that initiates thinking in the neighbourhood of the truth of Being. This for him is the ungraspable dimension of the Open, or "the Clearing"[173] where Being liberates us from subjectivity. He believes Rilke's poetry (unlike Hölderlin's) remains in the epoch of subjectivity and representation; holding on to the language of metaphysics and "word-forms", rather than being open to the unfolding of language from its very ground. I might suggest here that Rilke's Open is more like the opening up as it occurs in the threshold, whereas Heidegger's *Abrund* is that abyssal pure open that grounds as it withdraws. However, importantly for the discussion here, Heidegger also believes that Rilke's return to the Open is a radical *interiorisation*, a 'folding back' to the absolute heart of the subject. In other words, it becomes a kind of supreme subjectivity that experiences a contracted *personal* 'sublime'.

Yet, I share Michel Haar's reservations about Heidegger's belief that Rilke protracts the history of metaphysics. Haar maintains that Rilke's poetry instead celebrates and acts out the non-historical and enduring dimension of Being as it throws its light onto beings[174]. Moreover, he argues that Rilke's poetic conversion to the Open is *not* to the absolute heart of the subjective interiority, but is rather to, what he calls, the "pure space of the heart of the world". Therefore to leap onto the threshold is not a retreat to an "inner universe"; it does not mean we can only be at home within our interiority, as a kind of mystic without a world. Its experience cannot be a contraction of the vastness of Being into the content of the consciousness of the noumenal self. As the Open or the Whole is not reducible to human interiority, there must be a *two-way* movement where there is a transposition and exposition of the interior to the exterior. The subjective heart then becomes what Haar calls the "world-heart", which is neither

---

172.CP p.265.

173.Haar, Michel. *Heidegger and the Essence of Man*, op. cit., p.186.

174.SE Ch.VIII.

subjective nor objective. He insists therefore that in Rilke's poetry the Open is not thereby absorbed into the subject, but rather that, in this two-way movement, both the subjective heart and the Open undergo a "*transmutation*"[175].

In this process, where human experience overflows the realm of the already-known, there is an increase of value in itself, outside available measure. Haar writes that turning towards the stillness of Being, and abandoned of one's senses and presuppositions, the non-possessed heart "increases without name". In this expansion of the heart the subjective self gains the transformed world-heart through a loss of self-certainty and self-possession. The 'spreading out' or 'increasing' of the heart corresponds to the 'overflowing' of everyday subjectivity; a course needed for something to be shown in its essence, beyond its form, objectivity or utility. Being then discloses things *in a different light*. It is the sublime offering as excess, which spills over when the limit of things is reached. It guides one into the threshold so that the un-representable '*something more*' may be sensed. As the offering overflows the limit of representation and enables *communication* with the 'pure space', both the subjective heart and the Open are transmuted beyond representable imagination. That is why the increase is "without name"; what is happening here cannot be simply labelled or spelled out. And that is why Heidegger believes that the mediatory role of poetry is particularly suited to the 'art of naming' this nameless. The special kind of naming that poetry or art may facilitate does not define a simple presence, but enables an *experience* of thinking that simultaneously engages and exceeds the limit of representation. In the threshold, thus self-displaced and self-exceeded, the "innermost world-heart" communicates with the Open, becoming the transmuted heart beyond subjectivity.

Therefore, the increase of the heart does not mean the extension of the ego of the self. Indeed, it is the opposite: as Nancy writes, "when the sublime imagination touches the limit... it feels its own powerlessness"[176]. One is now near the abyss, where, in open vulnerability and in the presence of the elemental-being, objective and subjective beings are no longer the centre of attention. Yet there, in communicating with the Open beyond the ego, we may have a *sense* of expansion and liberation; an awareness that spills over the discourse of representation and metaphysics. The experience of the sublime is neither personal nor impersonal, but may be described as transpersonal or "suprapersonal"[177]. For Haar this is the experience of the transformed world-heart. So, although the sense of 'the sublime' I am probing here may be a kind of feeling, this cannot be taken in the commonplace sense. It is rather the displaced emotion that is outside subjectivity, yet still 'touching', and touched by, the limit of things within a *union* of the unlimited. The limits themselves, as the imagination touches

---

175. SE p.130.

176. FT p.233.

177. As does the theologian Hans Küng. See PA, p.248.

them, become in a sense semi-permeable or 'translucent'. Here we perceive 'something' of the truth of Being. This takes place in the threshold where the abyss, or the sublime void, may be perceived as the "totality of the un-limited" and experienced in the "self-overflowing of the imagination"[178]. The imagination, having touched the limit of presentation, reaches beyond itself to experience the delimiting by the unlimited itself.

Here, a region of sight is upheld that is 'open' beyond the senses. The unfolding of originary language occurs in the threshold precinct, wherein the reversal from the region of objects and of their representation into the innermost of the heart's space is realised. The deepest, hidden heart-space is not subjective human interiority, but corresponds and 'interplays' with the silence of the originary language of the unspoken "Voice of Being". As we near the pure open of the abyss and participate in its unfolding self-dis-closure, the secret, withdrawing dimension of Being "shines by way of be-ings"[179]. The creative determinations and expressions by human beings then 'come-to-pass', as the coming-passing of an *event* of deep existential significance, i.e. as *Ereignis* itself.

So this is an experience taking place within a different realm from the everyday; one where the Open is no longer barred and the abyss of its en-tirety begins to be grasped. The experience is permitted to rise, as the re-strictions to the Open dissolve. So, should this be interpreted exactly the opposite way to the radical interiorisation Heidegger perceived in Rilke's poetry: i.e. an expansion of the self out of the ego, whereby one's whole be-ing flows into the totality of Nature? This is how Heideggerian scholar Juli-an Young interprets the 'feeling of the sublime'; an expansion of one's be-ing into the infinite or the totality of all things, when the ego of the self has been surrendered[180]. I do not quite suggest such a movement with its cos-mic allusions and its implied dismissal of 'the ordinary' either. Moreover, its experience does not incapacitate reflective questioning; the threshold is not a trance-like state. Indeed, the experience is enabled at the very mo-ment that Being is retrieved out of its forgotten state and again becomes questionable. Although it is true that the self-centred ego has been put aside for the moment, Young's interpretation seems to imply a dissolution of the deficient self to become 'completed' in 'cosmic oneness with the uni-verse'. It is too easy to construe this as a kind of ultimate 'new age' experi-ence. The use of descriptions such as, 'expansion into the infinity of the cosmos' or "movement of cosmic love", could unintentionally suggest naïve interpretations. This expansion of the self appears to cling to a centrality of the human being, but now 'absorbed' in Being. His suggestion that in the experience of the sublime one "gives one's heart to Being", one is "nothing", a "mere incident in Being's venture" and "merely its conduit" as

---

178. FT pp.227,229.

179. SE p.133.

180. Young, Julian, 'Death and Transfiguration: Kant, Schopenhauer and Heidegger on the Sublime', *Inquiry*, vol.48, No.2, 131-144, April 2005.

the "only agent is Being itself", seems to retain a master/slave relationship. Such 'abandonment to Being' does not actually take hold of the unique mutual 'transmuted' relationship I am attempting to elucidate. For instance, it overlooks the 'need of the gods' for a clearing, an openness that can only be provided by an authentic human being. To no longer *own* the experience of Being does not mean we are *possessed* by Being. My analysis diverges somewhat from such total abandonment to the 'agency of Being' Although Heidegger at times seems to imply such a centrality of Being, in *Beiträge* this is not a 'owning' by Being of human being; as in the possessive owning of 'something'. *Ereignis* rather indicates an enabling of a non-possessive mutual 'owning', or *belonging*, whereby an authentic fundamental relationship is retrieved[181]. I suggest that one's heart, one's innermost way of being, is *transformed* in this experience. This transformation, involving the two-way movement with the Open referred to above, includes both the heart of the 'I' as this particular self, and the transformed non-subjective/objective "innermost world-heart".

When the "innermost heart" of the subject is transmuted to become the "world-heart" it celebrates a *unity* that is not entrenched in a stance that forces the exterior against the interior. Here all the accumulated baggage and hubris of power and knowledge are reduced to shadows without substance. Then, no longer holding onto an inauthentic home, a between is drawn near where, in the regenerating 'interplay' between the Open of Being and the openness of human beings, dwelling becomes a 'poetic habitation'. In the '*in-between*' of the crossing we *find* a grounded self for earthly dwelling. Moreover, this interplay comes to rest on the *Earth*, which is not only the sustaining and sheltering earth of physical matter, but also holds fast the ground of human dwelling. The latter is "that source, that origin from where the arising brings back and shelters everything that arises without violation"[182]. To perceive this emergence from the ground of dwelling things are 'let-be' in their essence; revealed yet sheltered, i.e. unviolated. Rilke, calling this origin the Ur-ground, alludes to its undefinable and overlooked nature when describing it as the "*unheard-of-centre*"[183]; the forgotten open originary ground that attracts and gathers. The silent voice of Being is "unheard" yet the stillness of the pure place of this centre gathers beings for earthly dwelling. These are then perceived and revered *as* the beings they truly *are*; i.e. marked by Being, not merely as objects faced by a subject, but by the being whose 'interior space', as the site for the moment, is 'open' for an event of consequence, and who is thus able to participate creatively in its growth and movement.

This affirms that at its roots the authentic or proper dignity of the human being does not depend on any subjective evaluation. In the transmutative pathway the subject is not obliterated; it is not as though the 'I'

---

181. Cf. CP p.xx.

182. PLT p.41.

183. PLT p.102, SE p.129

does not matter, but rather finds its dwelling place. What it discovers is not its property, but the heart's belonging to the Earth, the ground of human dwelling which in turn belongs to the heart. Heidegger, repeatedly, overthrows the human illusion that seeks ownership of the impossible: the possession of the essence of power, of language, of subjectivity, and of a metaphysical ground. Such an illusory quest fails to see the determinative thrust that these cast over our lives.

I have stressed that we are not 'important' as understood in today's restless corporate and instrumental world. The thinker, the poet, the creative person is also not important in that sense, which is merely one of prestige or impressiveness. Unfortunately, in the everyday the overriding desire for mastery seems oblivious to the fundamental human dignity. Yet, as Heidegger writes, "man does not decide whether and how beings appear". Instead "the ones to come" are to be "shepherds of Being". This means they are guardians of the Open, and as mediators participate in the ongoing course of its opening-up. The 'shepherding' does not involve a role of instrumental leadership, but it *does* require a manner of conducting oneself in such a way that "beings might appear in the light of Being"[184]. Herein lays the dignity and significance of human beings, as it is only these beings, capable of reflection and *poiesis*, who are able to participate in the interplay with 'the gods'. *Poiesis*, from the Greek "bringing forth" is *the act of creative founding* in a realm where the limits of pure objectivity and representation are no longer clearly defined. This creativity, as *poiesis* or poetics, encounters the other kind of truth, responds to it, and enacts it in creative engagement. I remarked earlier that genuine creativity is involuntary; it brings into the open that which is already gifted. Such poetic expression is a work of the imagination; yet not one that is calculative, causal-theoretical and explanatory, or a simple mental representation. It rather involves an imagination that on reaching a limit perceives an event of consequence that is beyond this horizon and becomes a fitting, indeed faithful, response; an openness to the *gifting* itself of a founding kind of truth.

So Here in the threshold, dwelling close to the *origin*, the subject has "moved" to be the transformed "subject of the sublime offering"[185]. Nancy writes that this offering does not offer the *Whole*... the present totality of the unlimited"[186]. As it arises from *Abgrund*, the "Whole" cannot be grasped, yet the offering invites ongoing reflection in a relationship of giving and receiving. This is a thoughtful experience that recognises the *significance* of the structure of receptivity, the 'space' of the sublime offering, signified in the image of the crossing of the threshold. What is being offered is the gesture of Being itself. Heidegger regards *all the unifications* that join the interplaying beginnings as grounding characteristics of the sublime offering. When we reflect on their features we may sense that they

---

184. FT p.184.
185. FT p.233.
186. FT p.238.

are all part of the richness of the gifting that occurs in the crossing, whereby the essence of things 'shines' for human perception. As offering, and in keeping with the transitory and transitional nature of the crossing, it 'comes to pass'; it is not a constant feature of life.

As observed, presentation and the un-representable are always intertwined. In the *interplay* between the beginnings, between the simple beholding of Nature and the grasping of its truth, they cannot be simply separated. Furthermore, post-metaphysical thinking has its own limits. To be sure, at times Heidegger seems to yearn for an unattainable untainted way of thinking, outside and prior to subjective, representational, systemic and metaphysical philosophy. Yet, he also affirms that our task is to simply become *prepared* for the *necessity* of this thoughtful questioning, which abides and sustains the unexplainable[187]. Such preparation cannot take place *in* the "first beginning", nor *in* the "other beginning"; it can only take place within a kind of 'in-between'; a *threshold*, which already suggests that a state of purity in thinking cannot be achieved. We cannot simply arrive at a destination called 'pure place'. In the elemental place of the threshold we do not purify ourselves from 'presence' and everyday experience; in any case, that would be an evasion of the difficulties inherent in being within the region of the limit and the 'strife' of the threshold. Yet, when the striving in terms of objects, logic and achievements is suspended and no longer a project, thinking 'touches' and overflows the limit. Representations then begin to fade, to be overtaken by the rising of a sublime presentation. This then is the *sublime experience*, taking place in the 'heart beyond subjectivity'

We have seen that this is an experience that exceeds everyday imagination. It takes place when the boundary of the customary is crossed to approach a realm of thinking that seems more open. At the limit of instrumental imagination thinking is transformed whereby the imagination reaches a maximum and 'self-overflows' the limit of representation into something that is no longer presentable, where it is "touched by...the union of the unlimited"[188]. Thinking is affected by the pure open of the abyss, the ground of Being. The threshold is a realm that has become more open because here the immeasurable openness of the abyss *encounters* the transformed opened-up imagination. When we are 'open' enough the unconstrained sublime itself overreaches the barriers that are usually placed in its way and is able to exceed the abyss towards the transformed imagination of the 'innermost heart'. The sublime offering is a gift that arises out of its inexhaustible ground. This threshold experience, in which the creative imagination takes centre-stage, nourishes the innermost heart. Here the hunger of the impoverished soul is satiated.

The sublime offering is a gift that may be received yet cannot become a possession. It is something to be received in a manner that does not alienate the essence of the gifting itself. So, in overcoming the instinctive drive

---

187.BQP p.148.

188.FT p.229.

for ownership we do not strive to 'own' it. Being is also not the owner of the gift, as Being itself is the gift. It is the happening of the gifting, the gesture or the granting of its truth, which gives the gift its ambiguous nature. As an offering that has as its origin a ground that withdraws it hints the way to an authentic manner of being, which is at the heart of becoming at home in a dwelling place for the human spirit. The fruitless fragmentation of intellect, of will, senses and beliefs is overcome by the gathering into the unity of this elemental experience. Once experienced, the decisions we must make at today's defining crossroad in history may yet be underpinned by a grounded wisdom.

So the experience of the sublime offering is not an experience of 'some thing', but one of an *act or gesture of giving*. The gift itself is the offering, which implies that the gift is 'given up', or "sacrificed"[189], to the one who is free to take it or leave it. It surrenders to the openness of human thinking. In this pathway the imagination is offered a less defined and more fundamental freedom that involves both the gift and the gesture itself. The sublime, which we may understand as simply another name for Being, is the 'pure open'. Although the open of the sublime can flow in us and be luminous, its disruptive and transformative capacity is not forced upon us. We have seen that to experience the sublime we need to learn to become attuned to the way it *hints* at 'something more', a deeper truth on which all things are dependent, despite these in technological thinking being perceived as absolute. The *call* to homecoming is gentle; its disruption occurs "without violence", via human desires, affinities and fascinations[190]. Therefore feelings need not be denigrated, but rather should be *grounded* through the transformed awareness taking place in the threshold. We are not to be too concerned with the emotion itself, but rather take this experience further in order to explore its *worth* to a sense of a life of consequence.

The sublime is not merely sensed in the beautiful or in that which overwhelms in its grandeur. As beings and 'ideas' are perceived in the overflowing at the threshold of the imagination, their forms are revealed from concealment as 'something more' than beautiful, utilitarian, etc. Here things are not simply grasped as pleasing, useful or otherwise. They are not immediately available for our 'pleasure', purpose, evaluation or dissection, such as when feelings, actions, forms and figures remain within our familiar everyday horizon. Instead their very essence comes into sight. Here we may find delight or a certain kind of serenity, but not a stagnant and insipid 'peacefulness'; the tensions at its limits displace everyday experience and false homeliness too much for that. Here we are near the open, 'beyond the pleasure principle'[191] and no longer held in bondage by it. This is the region that resists analysis and conceptualisation, where the boundaries of

---

189.FT p.237.

190.SE p.131.

191.Sigmund Freud, *Beyond the pleasure principle*, 1920.

angst, pleasure and joy are dissolved. The sublime calls and offers; it does not command to consent to the feelings of happiness, nor to melancholy or fear, but instead "gestures to pass beyond pathos and ethos"[192]. Here there is neither ethics nor aesthetics, neither logic nor irrationality; as we approach an originary ground such distinctions are defied in the threshold. The threshold becomes the domain for transformations in thinking to occur, when the offering is allowed to move and free us from the self-imposed self-enclosure of the instrumental and utilitarian mindset.

The realm of the sublime offering is truly one of deep significance, as it is here in the threshold 'beyond' the limit of everyday imagination that

> " ... *everything comes to pass, it is there that the totality of the unlimited plays itself out, as that which throws into mutual relief the two borders, external and internal, of all figures, adjoining them and separating them, delimiting and unlimiting the limit thus in a single gesture*" [193].

In this realm, where 'Being happens', the importance we attach to the value and the power of representation is suspended. These have reached a limit in their capacity to approach the sublime. The offering of truth traces a kind of 'contour'; a boundary beyond which we cannot sojourn, in case we plummet into the abyss and are overwhelmed by Being's excess. As noted earlier, we remain with the *thought* of the sublime and therefore the threshold is as far as we are meant to go. Yet, there is a "tension" at the boundary of the imagination, between the everyday and the threshold, where the limit is "stretched to the breaking point"[194]. Here between the boundary of something (this can be anything; a form or figure, a structure or system, a concept or theory, etc.) and its "un-limiting", there is the instant of rupture where its form is no longer quantifiable or definable. The sublime gesture and thinking, in their encounter in 'the heart beyond subjectivity' become involved in this "un-limitation" taking place on the boundary, on the border of presentation and at the threshold of imagination. The sublime offering enables the reaching into the 'open space' of the threshold that now seems more unlimited. Such "un-limitation" is not a 'some-thing', but 'arises' from its *ground*, as a de-limiting outline along the limited figure. By the insertion of something that is in essence held back (the light or truth of Being), here we understand that outward appearances attain 'something more' than simply an exterior manifestation. The unbound then bursts forth out from the pure open of Being and sensed as, what Nancy calls, "presentation in its *movement*"[195]. This is the emergence of Nature as *physis*. There the revealing/withdrawing, concealing truth of Being itself throws its light upon the limit of beings, whereby are both raised in their essence and effaced vis-à-vis their everyday instrumentality.

---

192. FT p.243.

193. FT p.229.

194. FT p.235.

195. FT p.226.

The 'goddess truth' comes to pass leaving mortal wanderers in the care of its primordial language. The threshold itself is not a hidden world, but once retrieved is an 'open space', a clearing where beings come to 'completion', beyond the limit imposed on them at the end of the first beginning. Beyond such a limit, beyond truth as correctness, the unthought and untrodden is not something that is lacking, but glimpsed as the "greatest gift" inherent in thinking, i.e. as the inexhaustible source for unique encounters[196] with a truth that reveals the very essence of beings.

The freedom of the offering gesture is directed towards the 'heart beyond subjectivity' and transforms imagination. We have seen that, as it has no owner, the gift of the sublime offering is not appropriable. Yet, as with any gift, a proper response is called for. But how do we find a fitting conduct to the gift of Being? It is a matter of being attentive, open to its reaching out for our innermost heart and to the out-flowing of its truth. It is a matter for being responsible for it and being engaged by it. It is a simple wakefulness when in the presence of other beings. And it is a sudden awakening that sees in wonder that a being '*is*'. It is to be a guardian who corresponds with and shelters the gesture of unveiling that which is concealed within it. It watches over the open clearing where beings can be seen in this light. This response may well be a moment of creative inspiration that becomes the seed for a surge in creativity. It is an experience that must be enacted, so that the response grows into an act of *poiesis*.

The artist derives her creativity and receives her role as 'one who reveals' from the sublime offering at the limit. From this boundary she begins to open up and *grasp* the very *ground* from which forms "are cut"[197], i.e. from where they "*once took their light from Being as the 'isness' of what is*"[198]. Taking place in the play of time and space, this ground then is involved in the limitation and un-limitation of forms, through its movement of, what Haar calls, "cutting and delineation". We know that concealment is the *precondition* for revealing[199]. Here emerging *physis* returns to *aletheia* to provide the surrounding 'medium' for the 'forms', now freed up for the creative mediatory task of the artist. "To create is to reveal", says Rousset[200], and to reveal is to uncover a limit, a boundary; to efface it and yet shelter it, and to show it *as* limit. As the sublime offering *withdraws* into concealment, its mystery, founded on the inexpressible abyss, is the lure for reflection and further creative engagement. Therefore, the artist's painting, the poet's words, the philosopher's ideas and the scientist' discovery, is at all

---

196. WCT p.76.

197. FT p.233.

198. PLT p.79.

199. This occurrence is described in Heideggerian fashion as "the emergent placing-itself-forth-into-the-limit", IM p.63.

200. In J. Derrida, *Writing and Difference,* trans. Alan Bass, London: Routledge & Kegan Paul, 1978, p.12.

times an *incomplete* revealing of its embryonic un-determinable source, which as the ground of Being always elicits new performative and poietic expressions. That is why fresh unique encounters with such works may always reveal further and deeper understandings and experiences.

We have seen that the sublime offering is a gift that may be received yet cannot become a *possession*. It is something to be received in a manner that does not alienate the essence of the *gifting* itself. So, in overcoming the instinctive drive for ownership we do not strive to 'own' it. Being is also not the owner of the gift, as Being itself *is* the gift. It is the *happening* of the gifting, the gesture or the granting of its truth, which gives the gift its ambiguous nature. As an offering that has as its origin a ground that withdraws it *hints* the way to an authentic manner of being, which is at the heart of becoming at home in a dwelling place for the human spirit. The fruitless fragmentation of intellect, of will, senses and beliefs is overcome by the gathering into the unity of this elemental experience. Once experienced, the decisions we must make at today's defining crossroad in history may yet be underpinned by a grounded wisdom.

While the open of the threshold is beyond representation, the experience of the sublime should not be regarded as a *general* one, that is, not as a nebulous sense of freedom or of otherness. The sublime experience is a moment of deep significance (that corresponds to *Ereignis*) and in the singular manner of the offering of its freedom cannot be of a general nature. Heidegger's "dwelling" is not some vague holistic feeling of being in harmony with creation. For him, it is something quite specific and yet undefinable. In the threshold a response has taken place to something 'non-general', i.e. the truth of Being that has been truly *grasped*. As remarked previously, in this encounter we are carried over, but do not get carried away in the turbulence - whether into confusion or into a cosmic sense of the supernatural or miraculous. To be carried over into the exquisite mystery of Being is to truly sense beings as wondrous, i.e. centres worthy of wonder, which is much more inspiring than anything supernatural. Thinking here is not remote and all-purpose; instead we have moved closer to what is already near, but has been forgotten. As human beings, able to reflect on such matters, we *belong* to the existential void, the open of the abyss. In the moments taking place in the threshold we are near to where we belong as human beings. There in the resonance of belonging, human beings and Being reach each other in their essence[201]. The fidelity to the relationship per se has been affirmed and sustained.

In the intimacy of this essential relationship with Being, truth unfolds in a manner that can no longer be simply 'aesthetic', or any straightforward representation of the imagination, but is illuminated by Being's previously hidden light. This experience is deeply primordial; it is elemental and originary. Yet, as noted, although able to frequent the site of the threshold again, we are not to linger there. The happening of the sublime offering in

---

201.Cf. ID p.37.

the threshold does not mean that one must languish in the feeling, but rather understand that one's heart, one's innermost way of being, is *transformed* in this experience. Unlike Peter's reaction, transfixed by the rapture of the Transfiguration, we are not to set up tents there. We are not to fix a mirage, neither are we to wholly abandon ourselves to the joy of sojourning the eclipse between representation and presentation. The desire to fix images of delight, to complete the exorbitant excess of the situation is an understandable attribute of thought and may be accepted as such. Feelings are not discarded in the seeking of significance, yet here they are no longer the driver and take a back-seat. Moreover, in the threshold one does not abandon oneself to be *overwhelmed* by the light of Being, but rather one *preserves its truth as a new kind of knowledge*. Again, to be *'carried over'* in the passage out of metaphysical and objective ways of thinking does not mean getting 'carried away'. Although we cannot remain there, after the storm of the threshold the peace of the site must be retained and the delight of being in the nearness of the mystery must be preserved; not consumed and discarded. The lingering stillness protects against doctrinal arrogance and the self-importance of prophetic proclamations.

Therefore, I cannot claim here to have "completed the [philosophical] attempt to understand the nature of the sublime"[202]. Heidegger insists on the impossibility of such a conclusion. When, in the transitionary thinking of the between of the crossing, we dare to attempt to communicate the inexpressible there can be no *calculating or declaration of finality*. The calculation of finality is inappropriate when applied to the inexpressible sublime. Its experience is not to be explained in the manner like an object revealing itself to scientific investigation. Heidegger also warns against the interpretation of this kind of questioning "psychologically". This, he writes, "...deprives philosophy precisely of the wondrous"[203]. Here we retrieve the wonder that dwells *in a between*, between the most usual, i.e. beings as objects, and their unusualness", which is their very being. It is *wonder* that first liberates this between and separates it out. It is specifically this 'between' or crossing that we have ventured and examined under the metaphor of the threshold.

Being, with its veiled manner of saying and showing, is to be pondered in a way that "has hardly been thought and [yet] is *not to be thought out to the end*"[204]. In the 'Postscript to "What is Metaphysics"' Heidegger writes "The obsession with ends confuses the clarity of the awe"[205]. Therefore, the mystery is not to be analysed to death, but instead sheltered and preserved *as* mystery; it is to be preserved because it is truly significant. We do not 'will' the unattainable into surrender. The gods, when besieged by

---

202. As Julian Young asserts, op. cit. p.142.

203. BQP pp.148,141.

204. OWL p.155.

205. M. Heidegger, *Pathmarks,* Cambridge: Cambridge University Press, 1998, p.237.

totalised instrumentality, can only take flight to the void of the abyss; they cannot capitulate. We cannot declare what the truth of Being *is* and yet we are its custodians. We can let its light shine on us; by thoughtfully reflecting on it, *something* of its truth opens up in the clearing we have created. We can stand before the mystery in the light of Being in awe and wonder, delight in it and seek deeper understandings. Heidegger writes about this light that longs to "open up" (the "need of the gods"!) and "illuminate" the inner hearts of human beings, so that they can perceive what is truly genuine in their dwelling places, in "their fields, towns and houses"[206].

In the pathway of being we have travelled we are interested where the thought of the sublime takes us. The heart beyond subjectivity, that does not seek to master or occupy, longs for the unity of Being and the human essence, the union of Being and articulated truth. The personal subjective heart of the everyday self has become, what Rilke calls a "distant", "strange", "far-off heart"[207]; a wanderer at home in an uncanny homelessness.

Wandering Companions of the Threshold - Soul and Spirit

Let us now return this strange innermost heart to the more familiar name of 'the soul', perhaps now seen in a new light. This soul is ridiculed by those who lack the poetic imagination and the humility to grasp its meaning. Or it may become forgotten by indifference, the shrug of the shoulder that comes with muttering "whatever". Often the ecosystem of the soul is made very unwell by the mistreatment of anxiety, ugliness, greed and brutality. Yet, the soul persists as the *strange* enduring human characteristic that enables us to step beyond the objective and the subjective, and reflect. It is attracted to, and concerned with, Being itself. Let us consider its 'strangeness'.

As we followed the above pathway of thinking we glimpsed the soul as the transformed subjective heart that has relinquished wilfulness. Thereby it has become "the stranger [who] goes under into the *beginning* of its wandering"[208], writes Heidegger with allusions of a movement toward the truth of the 'other beginning'. By overflowing the boundary of everyday objectivity, and thereby approaching the 'otherness' of the other beginning, the soul seems 'distant' and 'strange'.

The poet Trakl refers to its strangeness in *Springtime of the Soul*:

"Something strange is the soul on the earth"[209].

---

206. EHP p.37.

207. SE p.125.

208. OWL p.172.

209. OWL p.162ff.

Heidegger, contemplating these words, points out that here the "strange" does not indicate some isolated strange thing that inhabits our bodies as a ghostly presence and somehow survives our bodily demise. When taken in the context of the poem it rather contains allusions of movement of something, itself 'unhomely', towards an unfamiliar, yet essential, encounter that awaits it. The un-homely or the uncanny (*das Unheimliche*) is not to be understood in terms of an impression of fear or terror, but as the fundamental trait of the human condition that renders the soul as "something strange".

The soul is "something strange upon the earth" because it cannot fully inhabit the ground of human dwelling, or the pure Open of Being. Ultimately this means that "we belong to Being", and yet are not fully at home in this relationship. "We reside in the realm of Being and yet are not directly given access to it"[210]. As I noted earlier, the gesture of the sublime does not offer 'the Whole' and consequently human experience cannot grasp the pure open of the ground that is like an abyss. Yet, the continuing quest for meaning over the epochs of metaphysics confirms that the soul always seeks this ground of existence. However, this quest is too often in ways that are reluctant to leave behind the familiar; ways that are unable to let go of the known and consequently expect to find *a* ground. I have emphasised that this rather is to be a different kind of seeking, whereby the soul is 'set apart'. By moving from familiarity it nears a destination it cannot reach; hence the crossing is always 'a between'. We seek the evanescent 'Whole' knowing we can never find or see it. To accept mystery lets go of the ego's will for explanation. Yet it does not abandon the venture of questioning. Then, when mortals follow after the "something strange", they enter strangeness, writes Heidegger; and the one "called away" (by the echo of the gods) becomes "the stranger who is apart". To be sure, being underway towards authentic homeliness involves venturing towards a strange ground of dwelling that draws away as we approach it. Although in this sense we remain strangers to the fundamental ground of existence, it is not as the alienated banished outsider. The stranger (the one who has moved into this unfamiliar domain) is not imprisoned or exiled to bondage, but is the one whose essence is honoured and preserved. He or she has an awareness of always being underway towards home, while at the same time being rooted in the uncertain and enigmatic ground, which in Trakl's poem is called "Earth".

Heidegger reflects that the soul is on the way towards the Earth, the ground of human dwelling. This then is the essential nature of the 'soul': a wanderer, a seeker of Earth. Despite its abyssal nature, the soul does not *flee* the Earth, but instead *seeks* it and in this seeking the wandering soul's being is thereby fulfilled. In its setting apart the soul is sanctified. It must always be underway towards where its nature draws it. "The soul is called to

---

210. Malpas, Jeff. *Heidegger's Topology – Being, Place, World*, Cambridge: Massachusetts Institute of Technology, 2006, p.308.

go under" in its wandering; it is to release itself from the hold of subjectivity and objectification. In so doing the essential unity of humans and Being comes to fruition. In 'Language in the Poem' Heidegger writes that thereby the soul is carried into a strange domain of apartness, into "the farthest reaches of its essential being". Away from the homelessness of the everyday and responding to the reverberation of the gods brings about the apartness; the moment and the space of stillness needed wherein we can reflect, interpret, poetise and build creatively in faithfulness to being. This is how "only a god can save us". Thereby the innermost heart, in responding to the call of Being and illuminated by its light, becomes transformed. The soul has been recovered in fidelity to itself, and towards all the beings of the earth. The wandering soul, having been in the joyful presence of the gods and hearing the silent song of the 'spirit', is then able to "return more experienced" to earthly dwelling and to everyday life; a life that then cannot be conducted in a 'business as usual' manner.

Heidegger writes that "the *spirit* drives the soul to get underway to where it leads the way". What is the 'spirit'? The spirit is the light of Being, which *calls forth* the wandering soul. The strange soul follows the spirit, as the "flame" which "startles" and "lightens and calls forth radiance". In the threshold the soul then shelters and sustains the spirit; it *"serves the flame of the spirit"*, out of which 'inspiration' arises. At the same time the soul is the gift, the 'sublime offering' of the spirit. Here again we may sense the mutual belonging that embodies the shared need of humans and gods. The spirit is 'freed' to 'essence' in the unison of this relationship, where it discloses all things as "beautifully gathered"[211] in their most authentic nature. The spirit leads the way for those who have responded. They are the wanderers whose poetic imagination leaves them 'open' enough to give refuge to the strange soul in the apartness of the threshold. The spirit, as the light of Being for those who can *behold* its glow, guides the soul to get on its way, illuminating its wandering pathway. We do not know where this journey may take it and what hidden thoughts and authentic truth may be uncovered; yet therein lays the enchantment of the treasure.

To follow the spirit towards the essential of 'the Open', the unknowable ground of Being, one may have to plunge into the night of the spiritual twilight, into the very crossing of 'the highest threshold' close to the abyss. Here, as the stranger moves into the beginning of its wandering, it "goes under"; the soul "slips away" into the "shelter of stillness" of the void, near to where the gods rest. The safeguarding of the dark is not shunned by the one who knows both the gift of being and the dark night of the soul. In this journey towards one's essence, the wanderer does not attempt to exclude the strange apartness of the originary "Evening Land"[212]. In Trakl's poem the Evening Land expresses the passage into "the stranger's land that leads through the ghostly twilight" that belongs to the night; descriptions

---

211.ET p.143.

212.OWL p.194.

indicative of the crossing of the threshold on the brink of the abyss. This realm of 'not-being' retrieves the astonishment of being, so lacking today. The stranger that is the soul leads the way into this passage and *welcomes* the harsh retreat of familiarity.

"Night brings nourishment; the sun refines what's nourished

...

Night takes upon herself the renewal of our mystery; she robes the elect."[213]

This strange course can only be undertaken by a wanderer who, unconcerned about the customary artificial sense of security, is prepared to experience the silence and vulnerability of the abyss. The wanderer needs to be prepared - 'ripe' and ready for both the turmoil and the silence of the threshold. As one becomes more accustomed to sojourning the edge of the void, the fear and anxiety of the menacing and the terrifying (death as ultimate loss) is transcended. The "dark limit" of Death is then that horizon, beyond which is a mystery we cannot imagine[214]. It is not merely the end of our lives, but, in the effacing of customary ideas of sequential time, it is also the opening up of the space from which life is lived; from which it emerges into the wonder of a world illuminated by Being. Death belongs to our very being. In the letting-go and the awe of the threshold, change and death seems natural and astonishing[215]. Although the break with the everyday is unsettling, once experienced, the anguish of groundless homelessness and the numbing fear of the void are transformed in a way whereby life cannot ever contain the hollowness of nihilism. Now the void of finitude is not feared as an emptiness, which must be filled up with the contrivances of religion, the reassurance of the gurus of packaged bliss, or the fleeting happiness of addictive consumerism.

Along the pathway the wanderer receives sustenance from the refreshments found in the poetics of thinkers and others who creatively express the overflowing of the imagination. These are 'the friends' to which I return in the next section. We can find another metaphorical image of the wanderer in Heidegger's fine and original interpretation of primordial Greek thinking in *Parmenides*. Here we discover a "river" which has parallels with the image of the threshold. The "river" is not a feature of the landscape, but embodies the unity of a dwelling place, and a journey of becoming at home[216]. The river holds the "un-containable water" that the thinker must drink to make the crossing and nourish his or her relationship with

---

213. René Char, *This Smoke that Carried Us,* trans. Susanne Dubroff, New York: White Pine Press, 2004, p.119.

214. Malpas, op. cit., p.273.

215. Indeed, herein lays the possibility of a resurrection of the fundamental nature of death. Cf. Jean-Luc Nancy, *Dis-enclosure – The Deconstruction of Christianity,* New York: Fordham University Press, 2008.

216. Pöggeler, Otto. *Martin Heidegger's Path of Thinking,* New York: Humanity Books, 1991, pp.179-80.

the 'source'[217]. Its "water" cannot be contained, because its source is the *abyss*, the source of the uncanny; the origin of unhomeliness. Yet, the wanderer does not flee the ostensible emptiness of the source that seems to withdraw into nothingness. Therefore in *Parmenides* the river is called "*Carefree*"; here one no longer *fears* the 'nothingness' of the abyss. The wanderer has to 'stop' when near the abyss and 'attend' to it. In life's journey there must be such moments when we are truly attentive to the abyssal void. There has to be an *interruption* of the familiar followed by a thoughtful reflection. The wandering has gathered us here on an Earthly place, the ground of human dwelling. This place on the banks of the river "*Carefree*" is illuminated by the gods and the wanderer is thereby transformed. The "source" can only be approached by the one who has experienced "the river" and is thereby refreshed. Its "water" is a *gift* that nurtures the human essence and makes homecoming possible.

Earlier I referred to the primordial language that 'speaks' out of the truth of Being as 'the Greatest Text'. This essence of language resides here on the banks of the river close to the origin; it is where it 'comes to rest' as the 'house of Being'. Being 'essences' by way of this primordial language. When we entrust ourselves to wandering, a clearing is discovered on the margins of the river where the potentiality of language comes to pass and where it gestures towards the irreducible otherness of Being. Here in our explorations this language may be 'heard' and articulated in poetic and other creative expressions. Or it may simply be reflected upon, by those who have paused in their wandering to cross the threshold of the house and attend to its unspoken text. When this deeply concealed origin of language resonates in the clearing by means of human 'openness', the poem, or any response aroused by the light of Being, performs its 'saying' and its 'showing'. The Greatest Text is in turn sustained, and nourished by, the *difference* in the threshold. Because the threshold is an in-between, i.e. between objective everydayness and the source, thinking takes place within different spheres of influence; it is in transition and is articulated as such. The Text 'needs' this articulation; this is how gods come to pass. Divergent domains of meaning and involvement meet and interplay over the fissure created by the crossing. The articulation mediates between these worlds and enriches them by the poetic unveiling of their essential meaning.

Here, in the threshold near to the source that we seek but cannot possess, the 'Text' is heard as a silent voice, whereby the "light of Being" may inspire like a "lightning flash". Michel Haar recalls this hearing of the original delight of being close to this "voice", where within its grounded inhabitation "the smoke of the threshold" may arise within us "like the sacred spirit of wine"[218]. Although, in *Beiträge*, Heidegger uses the language of a 'thinker' rather than a poet his delight is clear when he senses the surpassing of the beginnings of metaphysics, allowing their playful interaction and

---

217. P p.119ff.

218. SE pp.154ff.

retrieving the primordial sense of truth, and of awe and wonder. In *Elucidations of Hölderlin's Poetry*, where his expression is more poetic, he writes that, "When the holy ray strikes... the poet's soul 'quakes'" under "the blaze of the heavenly fire"[219]. When thinking in the silence of the clearing corresponds with the unspoken Text, both the lack of words and the need for words is experienced. Then a more foundational language is heard and whispered out aloud by those who allow themselves the moment to hear. At times, when hearing the song of the spirit or seeing the light of Being there are moments when we cannot speak a word, yet the poetic imagination cries out for expression.

The image of 'the Sea' as the "Great Text", that primordial language out of which poetic thinking and articulated language arises, is also apt in these excursions of the wandering soul. Haar taking up its symbolism from Perse's poetry refers to the Sea as the "most elemental site"[220]. Alain Badiou in *Being and Event* similarly refers to the "oceanic site". From this inexhaustible source originary truth unfolds. The threshold resembles the limit of the shore of the Sea, and when one is exiled there the proximity of its 'silent voice' is sensed. The shore (limit) as threshold is where the 'displaced' self has chosen to move decisively into another realm. This is a homecoming of fidelity; here in remembrance to such faithfulness we pass through its open space.

The shore as threshold and the Sea, as the 'oceanic site', are appropriate metaphors for such unfamiliar habitation on the edge of the void. There is something about the sea, as an elemental 'pure place', that seems to bring human beings within a sense of being in harmony with the natural world and the cycles of life. Often, the sea can simply be a place to flee everyday responsibilities and pressures; a location for the pursuit of light-hearted distractions. Yet, the experience may also be one of stepping into another dimension. At times, the sea represents a darkness, a fear of the unknown, of its 'otherness'. Then it seems that our only option is either to flee it, or to be its servant, as one who awaits and responds to its silence. Sometimes the sea seems to calm our basic, universal and inconsolable grief, the *Angst* of the void, the abyss of the unimaginable singularity of one's death. Such experience is indeed abyssal, permeated by loss, being 'thrown' between birth and death, yet, which may be transformed when understood as a liberating encounter with its 'open pure space' from which a world may now be experienced, disclosed, and participated in, more meaningfully. "The Sea takes away and gives memory"[221]. This means that Being can withdraw from our gaze making us vulnerable to being overcome by the technological attitude, with its apparent allure of activity, vitality, confidence and diversity. Yet, remembrance is possible. Once we have experienced the unspoken language of the Sea, what was simply unfamiliar or something to be

---

219.EHP pp.91, 98.

220.SE pp.150ff.

221.EHP p.164.

feared becomes something deeper: the strange homeliness of wanderers that seek to be near to their true dwelling place.

We need metaphors and symbols, images and allegories to support us in the venture of homecoming and guide us in wandering the un-locatable "uncanny district"[222] of the threshold, where the essence of Being illuminates that which truly matters. Yet, how far and how often should we wander?

Returning to *Parmenides* (p.121) we find that Heidegger writes that the thinker must drink the *just measure* of water from the river "*Carefree*", and moreover should not only drink *this* water. This means that one should neither abide in the threshold too long, nor "stagger around in the lived experience... objectified... empty nothingness" (p.119) of thoughtless everydayness. As noted earlier, one cannot languish on the river's bank; we should not expect to constantly experience the distinctive moments of being on the shore of the Sea, or in the threshold in the nearness of 'the Open'. Otherwise, we would be besieged, in a state of excessive sensitivity to the given-ness and the sublime of Being, whereby one might be regarded as being 'on a high'. Such situations are always balanced out by 'lows'. These are the characteristics of mere moods, which although here they might give us windows into the abyss of Being, may also leave us in exhaustion, in a state of Nietzschean excess, besieged by Being's possibilities and abyssal no-thingness. The wanderer would then "receive more from the gods than we can digest"[223]. Perhaps philosophers such as Nietzsche, artists like Vincent van Gogh and other intrepid thinkers as, for instance, John Ruskin were such wanderers. Again, one must drink the *right measure* of the river "*Carefree*". Likewise, the Light of Being, which has its correspondence with the "river", can overwhelm: "Excessive brightness drove the poet into darkness", says Heidegger of Hölderlin's "madness"[224]. Moreover, the 'overflowing' of the poetic imagination at the limit has the capacity to flood the senses into delusions of 'pure consciousness' or 'cosmic wisdom'. These, like drug-induced hallucinations, may well be mere illusions of clarity. The senses are not to be so confused by the water or by the light that we are unable to retain clear-headedness for the sublime experience and incapable of reflecting upon its meaning. *Wisdom* should indeed come to pass here, but always that of a mortal, fragile and fallible human being. No, it is better to understand what is happening here, so that the 'right measure' of water is received. Too much of the water is perhaps no better than a life exiled from an intimacy with the earthly and Earth; both are likely to make us ill.

However, this is not a matter of the insipid modern inclination to find 'a balance': the averaged-out mid-point to opposing positions that leaves the problem it is supposed to solve to flounder in a kind of limbo. It instead

---

222.P §6 &7.

223.EHP p.61.

224.EHP p.62.

underscores the need to attend to the water, to shelter the light and to sojourn the shore, and to think and rethink. Then we can remember such moments of deep insight and significance in fidelity. Such remembrance leaves open the possibility for other unique encounters that experience the overflowing of truth and imagination in the primordial 'time-space' of another threshold. In these thresholds of thinking there is a 'co-respondence' to the "great Text of Being", where in this intimate relationship we experience the ebbs and flows of the Sea, its timelessness and timeliness, its uniqueness yet its resemblance, its remoteness yet its nearness, its muteness yet its primacy. Once understood and experienced, 'something' of significance then always remains with us in the everyday events and demands, the ordinary and the necessary. Now, as conceptions of certainty and truth, homeliness and homelessness are transformed, meaninglessness can no longer be totalising and indifference becomes redundant.

Although now "the soul is gathered into rightness"[225], and in its unity with the spirit is thereby fulfilled, its journey does not come to completion. Again, there is no calculation of finality, no computation of 'the whole'. This is not because "Man is not God", as Hans Küng suggests[226]. The Whole is not graspable, whether by 'Man' or by a supreme being. Therefore the soul on the Earth will always be "something strange". Yet, the stranger, who has risked crossing the threshold, is always under way towards home. To truly dwell in what is one's own is to be *underway* toward the source, the ground of human dwelling. Such seeking comes nearer to something that will always be 'distant'. The wanderer prefers a way of being where he or she can remain a traveller, a venturer of thresholds, lest the uniqueness of otherness be dissolved to sameness. In the threshold this otherness of the multiplicity of life is experienced. Once 'more experienced' the soul is always *free* to follow the stranger, the spirit that illuminates new sites of encounter; the uncultivated clearings in which creative building can take place. To remain a wanderer does not mean one is always wandering. There are times to wander and to be attentive to what is revealed along the way. There are also times to rest; to simply be and know that one has touched a realm essential to an authentic life. This is a journey of homecoming to a true dwelling place; not 'a' place as such, but one where the innermost heart belongs to the origin, faithful to the human essence. This heart, the inner soul of human beings, is then safeguarded because the confusion, in which humans find themselves through their feelings and the prescribed conventions of the culture, is transformed into the calmness of the earthly dweller.

In the threshold we leave an inadequate and taken-for-granted home and yet draw near to its source; it brings us near the ground of human dwelling. This source maintains the very site of the everyday home to which we may then return in a more experienced way: now aware of the

---

225. OWL p.176.

226. Quoted in PA, p.253.

significance of worldly being. When we are open to its essence this world may be truly *encountered* in a sensitive way, participated in and disclosed creatively. When the imagination 'overflows' the everyday, it reveals possibilities for a more deeply engaged encounter with all the beings of the earth. The modern existential and metaphysical crisis is then abated. The abyss has neither been filled, nor obliterated. Yet, having sojourned the threshold, discerned its awe and wonder and understood its centrality in the search for meaning, we are grounded. We are enabled to exercise our fidelity to being authentically human: at home as earthly dwellers who care for this earth.

Chapter VI
The Ones to Come – Housefriends

How are we to undertake this journey of homecoming? Is it to remain a solitary voyage in which we have no companions? Who is there to support us as we think our way forwards in this seeking? The threshold is a venture for which we need 'friends'. "Where are the friends to be found" asks Heidegger[227]. Here, these are not the friends who are simply good company to be engaged in conversation, but those anticipatory escorts who will join him, and hearten him in a journey of homecoming to the essential place. This journey is not predetermined; the homecoming involves an understanding appropriate for a particular self. Yet, while this transformative dislodgment can only occur in one's own pathway of thinking, we cannot simply negotiate it by ourself; we need these enduring friends. Such friends are the signposts, pointing to the indistinct pathway, which encourage one to depart and wander. I have already referred to these as the poets and thinkers who, for Heidegger, articulate the unification of "the Ones to Come". His dream for a "people of poetry and of thought" was to come

---

227.EHP p.149ff.

about by "these futural ones". As they allude to a task for us today, let us have a closer look at these now. In *Beiträge* Heidegger depicts three group-ings that he sees as playing a part in the transformation necessary for a spir-itual progress of 'a people', heralding the 'other beginning' as an epoch in

Western history in which human beings grasp a genuinely grounded way of being.

The *first* who grasp this 'greatest need' are for him indeed "the few": those thinkers and poets who hear the call of Being and are to arduously prepare the path of grounding and the possibility of sheltering Being's truth. Responding to the resonance of Being in such a manner is uncom-mon. This is not because it is a privilege reserved for the few or it requires great skills, but rather that it involves an uncommon experience of exist-ence that is not appropriable as a possession. We are unaccustomed to such a strange offering and moreover are unsure about how to respond. We have seen that a proper response to this kind of offering is one that needs a certain preparation, reflection and the nurturing of an unfamiliar aware-ness. To venture the threshold entails a transformative learning of the kind, which "lays hold of the soul itself and transforms it in its entirety by first of all leading us to the place of our essential being and accustoming us to it"[228]. In these later sections on Homecoming we are trying to get closer to this place. This kind of thinking is rare indeed in the modern world, where 'efficiency' and 'results' rank supremely. It has become difficult for us because the technological character of the contemporary world bars the way to reflecting on the realm of our essential being. This book has aimed to open up the topology and the way of thinking of this 'place'. Not so that we can become familiar with the experience of the threshold, as that will always be 'uncanny' and unique; but rather to become accustomed to the *need* to sojourn this domain and to faithfully respond to its significance.

So are we all to become poets and philosophers; to simply increase the numbers of "the few" who bother themselves which such matters? Obvi-ously that is not practical and probably not particularly desirable, so let us have a further look at the 'Ones to come'. As we continue, there is much in this *manner* of thinking that can be taken up by anyone who understands the need not to drift too far from "the place of our essential being". We have seen that this place is beyond the boundary of the everyday, where the self-articulation of Being and the self-emergence of beings occur in a way that is difficult to define, control and to express. Yet it requires human activity in order to be 'apprehended'. Thoughtful, articulated human lan-guage is to respond to, and correspond with the primordial language that I have described as the 'great text of Being'. Analytical thinkers, although they can question the experience of a transformative move from 'everyday' thinking and its meaning, cannot themselves simply express it, because the sayable receives its establishment from what is not sayable (inexpressible). Such a domain cannot be articulated in the usual representational and

---

228.Heidegger drawing on Plato's allegory of the cave in ET.

rationalistic way; therefore it understandably becomes a province for the poets and the creative arts.

To be sure, for Heidegger it is only a particular kind of thinker and poet who, as "the few", are ready to 'go under' into the stillness of the gods' call. "The poets are, if they stand in their essence, *prophetic*"[229], says Heidegger. However, this does not mean that they are to proclaim divine revelation or foretell the future. In *Parmenides* (p.5) he points out that it is not the job of thinkers and poets to declare revelations, proclamations and inspirations on behalf of the divine. Instead they are to open up a space and time for the appearing of "the gods" and to guide others in the direction of earthly dwelling. They pass on the hinting of the "need of the gods" and *mediate* their coming to pass so that others may also participate in this realm.

We are reminded that language is for Heidegger "the house of Being", and the sheltering and articulation of truth occurs through it. Therefore the poet, in being ideally suited to mediate this language, is foremost amongst those for whom the hint of the gods befalls and who passes on this hinting. The poet does this in a particular kind of poetry, which is other than ornamental, entertaining or cultural expression, but rather one that shelters and mediates truth. It is such a poet who is cast out into the "between", the sacred realm "between gods and men"[230] of the threshold. The poet has a special role to venture the pathway and express this movement towards the ground of Being and its disclosure. The poet must be 'open' to the foundational ground of language, in order to be able to work creatively with it. Creating and revealing belong to each other. In order to reveal the poet needs to 'hear' the "great text of Being" and for this must make available a clearing (as a kind of opening of the imagination) necessary for the passing of the gods. For Heidegger, the poet, for whom the *essence of poetry itself* has become worthy of probing, is the one who "risks more" and is "more venturesome". Today, such a "poet in a destitute time" has to experience the defencelessness and the "unholy" of the "fugitive gods". He or she has to truly endure the modern malaise so as to be a prophetic mediatory voice of essential thinking in this epoch of the forgetting of Being. This poet has to undergo the distress that the absence of the gods brings about. Only then, in a response of fidelity can the poet be underway to the "integrity of the sphere of Being" and draw the god nearer[231]. The gods then are the ones who "brighten" something that the poet, in the transformed heart, then "brings to light" for understanding[232] in creative writings, thus fulfilling a mediatory role. Again, 'the gods' are not beings of any kind, but moments wherein something of deep significance emerges out of Being itself.

---

229. EHP p.136.

230. EHP p.64.

231. OWL p.139.

232. EHP p.39.

Poetry, as a thankful and a thoughtful response to such events of significance, mediates this bringing into light whereby things are perceived out of their very essence. Despite changing situations, fashions and attitudes the poetic *spirit* persists through the ages. It is offered to *all* those who know how to be silent enough so they can respond to the silence of the sacred realm; the mystery of the stillness where the gods rest. The poem that names Nature as *physis* did not arrive for the first time with the emergence of philosophy in ancient Greek times, with what Heidegger calls the 'first beginning'. The event of Being that is named in the first beginning and deepened in the other beginning has its poetic happenings in multiple, more ancient sites, such as in China and Eastern thought. The first Australians, in their own unique way, no doubt have sensed the emergence of Being for at least 40,000 years. In the Ngarinyin language they have a word "yorro-yorro", which means something like "the spirit in the land that makes everything stand up alive"[233]. I do not think it is pushing the analogy too far to suggest that this word names the same event as the Greek *Aletheia*; the unveiling, emerging, and bringing into life of Nature as such.

Therefore, the poetic spirit, as the Poem of Being, persists through the epochs. This gives the poem and poetry its timeless character; it has always echoed an enduring bond to the theme of nature. As discussed, Heidegger never meant the "beginnings" to be historical moments in chronological time. Likewise, the Poem maintains a non-epochal lasting presence beyond the reach of civilisations and traditions. Heidegger simply found much in primordial Greek writing that demonstrates his conception of the fullness and vitality of emerging nature and the event of originary truth.

All meditative thinking is poetic. As thinking and poetics both relate to language in an essential way there is to be a dialogue between them, so these can reveal its originary role[234]. We know that Heidegger struggled to overcome representation, yet nothing can appear before us wholly un-interpreted; language comes to us in some contextual setting. Articulated language is the outcome of both a gifting and a receiving that, if grasped together, shelters and expresses an essential yet fragile relationship. Poetising and thinking in their essence are both open to primordial truth[235], whereby originary remembrance springs from the *dialogue itself* between thinking and poetry. While poetic reflection and thoughtful questioning belong to each other and must remain *near*, Heidegger nevertheless suggests a distinction between them, as their allotted tasks are different. Poetry, through its thoughtful naming of gods and touching the essence and limits of things, is able to reach deep into the sacred ground that is "in the presence of the gods". The "poetic *spirit*" itself already "dwells at home in the grounding ground"[236]. We recall that the spirit, as the 'light of Being,' needs the soul; here it finds this essential relationship. The poet's

---

233. M. Leunig, op. cit., 'Our flagging enthusiasm', p.21.

234. OWL pp.136,160ff.

235. Cf. Pöggeler, op. cit., p.167.

creativity, held in reserve within the soul, must first of all be *at home* near this ground before mediation can take place. The poet then *names* the sacred in creative poetic expression.

However, the *thinker* asks the difficult questions about the most unfamiliar sphere of Being that the poet names. The thinker, in her reflections, makes what underlies the poet's words visible. In a more expositional role she has to reveal the essence of this questioning as an act of "remembrance" that honours both the "friend", i.e. the poetic spirit that endures, and the 'friendship'; the relationship itself with this spirit. Both thinker and poet have to be at home in this strange (uncanny) homeliness. The poet has to do the 'saying' of language, about which the thinker can again reflect. He thoughtfully addresses the reader, yet leaves unspoken the insight out of which he speaks. The thinker may attempt to articulate the unspoken, but can only do this in a mindfulness that reveres its poetic essence, i.e. not too 'objectively'. This relationship between poet and thinker nurtures the soul of anyone who knows what it means (and dares) to be going to the 'source'.

Yet it seems that for Heidegger only few poets and thinkers have initiated the grounding and sheltering of the truth of Being in this manner. Perhaps only Hölderlin fulfils all the requirements Heidegger prescribes for "the few" of "the Ones to Come". Whereas he felt that Trakl only took a path away from "old degenerate ways", Hölderlin's poetry *overcame* representational metaphysics and the prejudices of western thinking. He is seen as "the most futural of the ones to come"[237]. Hölderlin, in the true role of the poet, is the master at 'naming' the event of consequence (*Ereignis*). He is "the poet's poet", because his poems grasp the very essence of poetry itself, as "the thoughtful confrontation with the revelation of Being"[238]. In Hölderlin, or more precisely in the way he invokes the "poetic spirit", Heidegger finds the kind of "friend" he seeks. His poetic writing is foremost in exercising the mediatory task of communicating the nameless. It brings another kind of truth into the light for reflection and prepares a ground for genuine dwelling. It thereby paves the way for 'building', i.e. practical action of the kind that lets the earth *be* earth. When it comes to contemporary *thinkers*, "the Few and the Rare... who from time to time ask the question"[239], we might well wonder whether Heidegger would consider anyone but himself. He believes that although Nietzsche truly experienced the end of metaphysics, he did not succeed in '*overcoming*' metaphysics and the limits it imposed on itself. However, we should always remember that all human existence is thoughtful and poetic *in its essence*, and therefore

---

236.EHP pp.60,115.

237.CP p.281.

238.EHP pp.52,9.

239.CP §5.

these reflections apply to *all* seekers of the uncertain ground of existence, at the conjunction of the known and the unknown.

***

The *second* group of "the Ones to Come" that Heidegger names in *Beiträge* are "those many allied ones" who are to enact, and build on (by manifestations of "dwelling"), what "the few" have prepared. In his *Elucidations* he also refers to these as "the others"[240]; those thoughtful ones who learn to 'listen' patiently and carefully to the mystery and the stillness expressed in the poets' words. We again should keep in mind that the unification of the "Ones to Come" should not be constrained as an elitist vision. "The few" are not 'more important' than "the others". Hearing the echo of the call of Being and exceeding the everyday does not require some selective 'learnedness'. Self-importance is alien to threshold thinking. 'The few' simply affirms that some people, whether poets, artists, composers, thinkers, writers etc., have extraordinary gifts to articulate the inexpressible. Yet, it is not only selected poets and thinkers, who are summonsed to the awesome and disquieting mission of mediation between human beings and the holy, and who are called upon to respond to and shelter the call of the truth of Being.

To seek, preserve and shelter this numinous truth, central to homecoming, is for all. This journeying is never completed as it never really leaves the place in which we already are. However, now the sense of home is fundamentally reoriented towards the inexplicable source of wonder. This homecoming is the return to the nearness of Being. This is not just for an academic for whom the southern German landscape provided some of the images and language that expressed the poetics of his thinking. Certainly, Heidegger regards a dialogue with Hölderlin as being of particular significance for the German people, who, following in the footsteps of the Greek history of thought, embraced a technological world that needs to be transcended. While his perceived association between the ancient German and Greek languages is somewhat dubious, it is clear that "the treasure" that belongs to the German homeland is the foundation of *all* earthly habitation. It is the *same* treasure because it is found in response to the *same* call to homecoming. It may be recognised and celebrated in a myriad of diverse ways, but in its essence it belongs to all human beings. As noted in Chapter 1, homecoming to the treasure preserves what is most appropriate to the people of *all* homelands. *All* human beings who understand the bond to the mystery of the truth of Being are its "seekers", "preservers" and "guardians"; they are "the ones to come" who are 'open' and 'modest' enough to let events of deep significance happen. Homecoming contemplates the Holy; that centre of meaning that the poet invokes[241]. It seeks what is always already near, but is habitually forgotten. Indeed, it is nearer than the "well-known things and simple relations" of the everyday.

---

240. EHP p.47.

241. EHP p.222.

However because it is not very obvious and its demand is gentle, it is mostly overlooked and passed by. As the poet René Char writes:

"What is reality without the dislocating energy of poetry?"[242]

Obviously, not everyone can be a poet or philosopher in the 'official' sense. But if the great need of modern homelessness is to be addressed, the relevance of seeking a unifying and grounding domain is not just for poets such as Hölderlin, or for 'Heideggerians'. Here we are not necessarily concerned about putting pen to paper in a moment of inspiration, but more about recognising and awakening the poetic spirit that resides in the being that is human. At any rate, Heidegger in *Beiträge*[243] lists a number of ways that the sheltering of the truth of Being may be put into practice, such as thinking, poetising, building, producing and arranging works, guiding and leading, sacrificing, suffering and celebrating. As this fairly well covers the whole panorama of life, we can conclude that this is applies to all of human existence. However, life is now to be lived *awake* by those who "dwell-forth", aware of what it means to be. When Heidegger uses this term (*bewohn*) it refers to a genuine dwelling that implies a movement of seeking towards something specific, i.e. towards "the house of the world"; a seeking of Earth as human being's ongoing dwelling place. Creative expression and grounded dwelling is the role for all who are alert to the unique and inexhaustible manner in which all things emerge from their veiled meanings, and who seek to make this experience their own. Creativeness entails a way of thinking that shelters the truth of Being in all beings and allows 'strife' to happen in the wandering of thresholds. In the open space of the between, the free play of this rapport is acted out in creative and innovative work. Who knows what unique, unanticipated and astonishing happenings may come to pass when imagination grasps the tension that 'opens up' the sheltering, concealing Earth (as the ground of human dwelling) in relationship with the arising world of human making?

This 'opening up' may occur in all kinds of manners. Such poetic Earthly dwelling may be understood as integral to how we see our personal role in the world. For thinkers such as Heidegger free dwelling that transcends ordinary habitation occurs in the non-metaphysical neighbourhood of Being, which does not fall back on a God that is supposed to explain everything. The poet may look for this dwelling in 'earthly', more elemental places, the intimate, normally hidden gifts of nature. The scientist might seek out the forgotten awe and wonder of the uncovering and the yet to be uncovered, and the exquisiteness of the undiscoverable. The technologist or engineer might reflect on the pre-given structures and interrelatedness that make 'building' possible. The elegant, potent and profound laws of nature may well contain seeds of meaning that await a poet from unexpected realms who can create a narrative to rival any creation myth, with the added bonus

---

242. René Char, op. cit., p.123.

243. CP p.213.

of not being a fairy tale[244]. Poetic dwelling belongs to the essence of human beings, so it cannot be confined to those we call poets. It belongs to all "the others" who perceive its need in the face of the flight of the gods. All can be poets/thinkers when understood in this 'broad' sense.

These then are the 'house-friends', the friends of the house of Being and the house of the world. "To dwell poetically" is to again stand in the presence of the gods and to experience the originary, essential intimacy of things in a threshold moment of disruption and transformation. We have seen that this can only be a transitionary event, as the gods are not there for us like ready at hand items, prescribed and packaged for consumption like items on a supermarket shelf. Rather, they are encountered as an offering of the sublime, a gifted moment of insight into that which simply *is*; a moment where the "lightning flash of the god" illuminates a way of thinking and seeing that was previously held in darkness[245].

***

In Heidegger's dream, this transformation then flows out into the *third* grouping of "the Ones to Come". They are "those many who are interrelated by their common historical origins, through whom and for whom the recasting of beings, and the grounding of truth, achieves durability"[246]. This is sometimes interpreted as the more contentious aspect of Heidegger's vision; a dangerous ultra-nationalist movement of the '*Deutsche Volk*'.

Yet, as I have already pointed out, Heidegger's reflections do not resemble the rhetoric and even less the practice of the movement of National Socialism. The manner of Heidegger's quest for a different truth to everyday notions runs wholly counter to the common sense of the epoch and the public world of modern society. We should not place unwarranted emphasis on inserting Heidegger's thought into a political setting. As I have contended, the 'crossing over' to the other beginning involves an essentially personal mindfulness, yet is one that Heidegger hopes will flow out into the culture.

A sense of cultural and geographical belonging cannot be tribal if we are to know our true dwelling place. Fundamentalism suppresses uniqueness, preferring a stunted dream of the collective. When an inadequate view of culture confines, shackles and stifles thinking, an inevitable conflict arises between it and the universal nature of a common humanity that holds fundamental characteristics unique to human beings. The limited perspective assumes that a particular group or community is *the* positive expression of an idea of humanity. Yet it is one that only serves the perceived interests of that particular group. However, a genuine universal humanity is grounded in an essence (explored in this book) that is common and deeply significant to all human beings. When such significance is overlooked it restricts perceptions of culture, tradition and values. I suggest that deficient cultural

---

244. Ian McEwan in PA, p.360.

245. EHP pp.60ff.

246. CP p.67.

relationships between humans or societies have at its source a loss of this sense of a fundamental universal earthly ground. A more inclusive view of culture arises from understanding this essential ground, out of which values and practices can be derived appropriately. This awareness results in a thoughtful perception of culture within a view of worldly dwelling that is less influenced by the familiarity, reactivity and negativity of pre-suppositions and feelings. This perspective is not neutral or indifferent, but grounded on an enduring and holistic self-understanding that affirms the significance of a humanity that is both culturally diverse *and* universal.

Although Heidegger's talk in *Beiträge* about the growth arising out of the few to the many seems to suggest a mass movement, he is at pains to emphasise that he is referring to a non-quantitative and non-historical site and moment of *Ereignis*[247], as an event of deep significance where there is a kind of *growth in thoughtfulness*. Is it too much to hope for such growth in a world where the instrumental attitude holds sway? Heidegger asks "is there any greater fear today than that of thinking?"[248] The banality and the vacuousness of much of modern discourse seem to confirm the destitute times Heidegger already spoke of. And more dangerously, as Hannah Arendt declares in *Eichmann and the Holocaust*, sheer thoughtlessness expels one from reality and opens the way to absolute criminality. Thoughtfulness is not reactionary. Yet, perhaps the pressing nature of the social/environmental crisis will bring about more than instrumental thinking regarding survival methodologies. Maybe the hunger for the uncommon will be recognised as "a hunger in oneself to be more serious"[249]. Perhaps the question of what it means to be will again become questionable. This is always a matter of *thinking* and *rethinking*, within a moment where a more fundamental truth is able to emerge. It comes to pass in the clearing founded in the threshold where Being's light can illuminate *properly* considered responses to the unprecedented predicament into which we are cast today.

In the late essay *Hebel der Hausfreund* Heidegger concerns himself with another "friend", the early nineteenth century writer Hebel, who lived "in the bright nearness to language"[250]. Because he was 'open' to originary language Hebel is an "authentic housefriend" (*eigentliche Hausfreund*). As noted, for Heidegger, language itself is "the house of Being". It is thereby the primal home of the human essence. Hebel then is a "friend to the house of the world"; i.e. a friend to the ongoing dwelling place of human beings. When human beings are such friends, this world becomes a gathering that is indeed "a between", "between earth and heavens, between birth and death, joy and sorrow, between work and word" and between thinking and

---

247. As, what Heidegger calls, *Geschichte* rather than *Historie*

248. PLT p.77.

249. Philip Larkins in PA, p.210.

250. See the discussion in Lovitt, William and Harriet Brundage. *Modern Technology in the Heideggerian Perspective,* Vol. I & II, UK: The Edwin Mellen Press, 1995, p.646ff.

building. It is a between that must be inserted into the experience of life if it is to be consequential. The turmoil and interplay of the threshold (which plays out in a proper collaboration between human beings and the environing world) is thereby both a struggle and a delight. The gods can only appear when we become poetic wanderers such as these friends that "dwell within the most joyful"[251]. They gather what already belongs together; making the 'house' a true dwelling place.

The need persists to become aware that the habitual home of the everyday lacks such 'housefriends'. Heidegger writes, "...today we are erring in a house [dwelling place] of the world from which the Friend is absent"[252]. Modernity's dilemma is how we are to live with technology: Are we to be victims of technology (now more pressing then ever, both in an instrumental and in an existential sense), or are we to recover "friendship for what is essential, simple and lasting"? Clearly, expressions of 'house-friendship' may be found in the domains of the arts, literature, poetry, music and philosophy, as well as in the work of those whose care for the environment comes from something deeper than mere survival instinct. These should be sustained so that they can generate other mediatory friends who remind us of a fundamental ethics that considers and nourishes the dwelling place and the very essence of human beings. Such friends need not only be sought amongst a select few but may be encountered in the everyday; *if* we are attentive enough.

The unfamiliar kind of probing pursued in this book is uncommon in much environmental philosophy or ethics, where the matter of concern is centred on the individual perception of nature and the subsequent manipulation of human eco-psychology. Instead, I have emphasised a thinking that is both transformative and transformed, whereby subjectivity loosens its hold. The true worth of things beyond utility is sensed. The primordial human essence of 'care' for all of nature then infuses a way of being. When seeking the ground of human dwelling a deep care for all beings amongst which we find ourselves should indeed be the allotted outcome. Yet, I should note here that this transformation in the way we 'see' the earth and its beings does not automatically guarantee behavioural change. We know from our own experience that feelings, the ego, the desire for comfort, the "stranglehold of the fix"[253] are powerful things. No doubt constructivists, sociologists, psychologists, eco-philosophers, environmental scientists, etc. can also help us to make the necessary life-style changes. Yet, here we have glimpsed the possibility of a less destructive, much more deeply engaged encounter with all beings of the earth. We have found echoes, hints and possibilities for a less belligerent, self-protective and self-important living, and moreover for bringing about ways of thinking towards the retrieving

---

251. EHP p.36.

252. Quoted in SE p.91.

253. Frank Fisher, op. cit., p.4.

and sheltering of what is most precious in earthly dwelling and what in essence cannot, and must not, be objectified and used up.

Clearly, from the foregoing discussions, we cannot separate 'being at home in the natural world' from being at home in the world of human activity. However, having retrieved the significance of Being as dwellers of the earth, such activity is now manifested in a more care-full and attentive manner. It would be wrong to assert that Heidegger is 'an environmentalist' in the modern sense. Although he seems to eschew activist polemics he cannot be said to vacillate where he stands on Nature, Earth and earth in relation to 'the environment', when seen from the "broad" perception. When Heidegger speaks of a "broad" perspective, he does not mean in the sense of "extended, approximate and superficial, but rather in the sense of essential and of a grounding fullness"[254]. Elsewhere he describes such a perspective as "far-reaching, rich, containing that which has been deeply thought"[255]. Let us read his compelling self-dialogue in §155, 'Nature and Earth', of *Beiträge* (modified in my interpretation), which encapsulates some of the themes we have explored.

"What happens to Nature in the epoch of the overwhelming 'technological attitude' towards the world?

What happens to it when separated from beings by instrumental objectivity and examination?"

"Its growing and ultimate destruction".

This entails not only its demise as a vast, complex and astonishing ecological system, but also the obliteration of its very essence as the ground of human dwelling.

"What was Nature in its originary conception?"

"The site for the moment of the arrival and dwelling of gods, when it, as *physis* (the self-emerging source for awe and wonder), rested in the luminous essence of the abyss of Being. Since then *physis* has been demoted to a being[256] and thereby reduced to something that can be measured, instrumentalised and evaluated. Finally, all that is left is nature as a mere resource, or as scenery and recreational opportunities organised for mass production and mass consumption. Is this then the end of Nature?"

"Why does the Earth keep silent in this destruction?"

We have seen that it takes a special kind of 'hearing' to heed the silent voice and the hint of the need of Being; one that occurs via the poetic imagination that leaves behind strategic calculating. It entails a unique manner

---

254. Heidegger in *Pathmarks,* op.cit., p.190.

255. 'Anaximander's Saying' in *Off the Beaten Track (Holzwege),* Cambridge: Cambridge University Press, 2002, p.250.

256. An object, even if it is 'alive', which is only one particular characteristic of the beingness of beings, cf. CP p.194.

of 'seeing' to perceive the earth as Earth, the ground of human dwelling, which is the foundation for grasping the essence of the earth. Heidegger then reaffirms that the relationship between Earth and world is always one of 'strife', a necessary 'struggle' of emerging and disclosing, and of withdrawing and concealing.

> "Because Earth is not allowed the strife with a world and thereby not allowed the truth of Being."
>
> "Why not?"
>
> "Because the more gigantic man becomes, the smaller he also becomes".

This is the loss of the essence of human being (of self-being), arising out of the constant striving for self-importance, which restrains truth to instrumental correctness only. The suppression of this truth by "calculation", "acceleration" and "the outbreak of massiveness"[257] leads to its diminishment and ultimate oblivion. "The outbreak of massiveness" refers to the unlimited, self-aggrandising increase (calculative growth) of *all* things, such as in the semblance of 'important' events and projects presented for an obligatory requirement of 'admiration', resulting in the lack of understanding (indeed the overwhelming) of what is genuinely significant in beings.

> "Must nature be surrendered and abandoned to the utter loss of its essence in the fading of the last god?
>
> Are we still capable of seeking the Earth anew? Who enkindles the strife in which the Earth finds its Open, in which Earth encloses itself and *is* Earth?"

This is the challenge placed before us, as 'the Ones to Come' and as 'housefriends'; the *decision* made necessary in this unique moment in the history of human existence, where "the end of Nature" has become callused in the wounds inflicted by the dominant utilitarian paradigm. Today nature presses upon us the necessity of a decision of how we are to live; we are truly at the crossroads. Are we to subdue nature to such a degree that its essence can no longer emerge for our gaze? When thus suppressed its resurgence may well be in a manner that will make things very uncomfortable for the self-indulgent culture. Yet, the ecological crisis offers the possibility of a genuine relationship of human beings and the earth, grounded in Earth. The fundamental understanding of what it means to be human explored in this book is the unshackling and enabling framework necessary to make the authentically grounded, ethical, intellectual and technological decisions that are so crucial today. To save the earth is to release its essence, to 'let it be' earth, as the 'web of life', *and* Earth, the ground of human dwelling. Those who struggle to preserve the ecological integrity of the earth and bring about genuine sustainability might reflect on *what calls* for a care for the earth and *what it is* that is being opposed. The case for the

---

257. §56 and §58 of CP.

protection of biodiversity is often propped up by appeals to the value of the *services* that healthy and resilient ecosystems provide for human well-being. They are deemed valuable, i.e. measurable, calculable, assessable, in terms of human food security, water quality and supply, erosion control, health, food production and security, air quality, pollution absorption, tourism and recreation, etc. Of course, all these are genuine benefits, but where is the reverence; the awe and wonder at the mystery of Nature? Where is the faithfulness to the intimate relationship with the garden into which we are cast?

If we are to merely respond to the environmental malaise in the terms demanded by the instrumental attitude do we not, in a sense, prolong 'more of the same'? Are we not simply evaluating and calculating which of the abstracted elements of mechanised models of nature are 'important' or 'unimportant' for our utilitarian purposes? An authentic response clears the blockages that impede a change in our inner way of being. The decline in human relations with the earth, by clinging to inappropriate practices in an anxious attempt to maintain unsustainable lifestyles, cannot be simply surmounted by technology 'saving us from the danger of extinction', but by saving the very essence of what is possible and what is not in human thought, action and experience. Human beings need to learn not to overstep, but dwell in, their possibilities. This preserves the quiet law of the earth in its self-sufficiency, where all things emerge and perish in their allotted spheres of possibility[258].

To see life afresh means attending to the wandering human soul; that which waits in reserve to emerge from its forgotten or abused state to again make known its fruits of truth, hope, care and beauty. Transcending the everyday selves in the threshold is an experience that can be valued most when its worth is understood; such significance is what ultimately truly matters in the grounding of the human-nature relationship. Then we always stand before the decision between the end of a first beginning, now ossified and objectified, and another beginning; one that 'frees' the first and enables the return of the strange human homeliness to which we belong.

Concluding Wanderings

The insights that have come to us as hints and echoes do not readily accommodate themselves as in the dot points of the Executive Summary. If I can summarise them at all I would suggest that they rekindle a thoughtful and deeply significant poetic imagination that lies within all human beings. The awakening, reclaiming, re-visioning and enacting of this imagination get us ready for homecoming to what is one's own; close to the source of the sacred and the spiritual.

---

258. Cf. EP p.109.

The thinkers and poets I have briefly engaged all deepen the understanding of the threshold experience. Although the details, their derived perspectives, may differ, their deepest insights do not contradict. They should not even be seen as presenting *alternatives*, as their various 'translations', transformations and moments of insight gather upon the dwelling ground of human existence and affirm its un-representable, un-thought dimension. Pure and unmediated access to some 'transcendent other' is unattainable. I am not claiming a pathway to an exalted mystical state of being, but to a way of being that is more worthy of the being that is human. There will always be moments when language will be appropriated back into its metaphysical character; where words will yet again become mere words making up propositions about given objects, and when the deep experience of performative thinking in the threshold seems to have faded. Yet, once the significance, indeed the sacredness, of the event has been experienced, it will no longer *remain* out of sight and forgotten.

We have seen that in the threshold, where the imagination exceeds itself, we do not find the kind of homeliness in which we can be comfortable. Its margins are too effaced, too ungraspable, and its nearness to the fathomless of the abyss is too disconcerting for such unperturbed being. Away from the familiar realm of the everyday, being in the threshold entails a state of exile. Recovering the enigmatic proximity and yet distance of both the familiar and the unfamiliar finds one in a kind of banishment from the everyday place where only the customary limited kind of truth as correctness is acceptable. It is not only the exile from the everyday that unsettles us here, but here we are also confronted with the 'strangeness' of the human soul. It brings us face to face with its homelessness. Here we come to terms with our deepest nature and need: seekers of our earthly ground. Hence to become at home can only be via a "not-being-at-home"[259]. The unfamiliar threshold must be engaged for genuine earthly dwelling. Its seeking is an uncertain, modest course of cultivating one's roots to the *source*, the dwelling ground of Earth. The pathway that needs to be trod is a movement of opening up to the primordial 'great text', as the language of Being that arises from the need of the gods. We may then mediate and bring to language the poetic dimension of the earth. When this dimension is exclusively substituted by structural and conceptually transmittable categories, i.e. by objectification, something is lost that is less tangible yet of deep significance to human existence. Only once we have become 'near' to the Earth may we poetically dwell upon it. Then we may genuinely 'build' in ways that respect the being of the earth and its limits and thus save the earth *as* Earth. The earth is thereby not only saved as the human 'spaceship', a vehicle for our survival, but saved in its very *essence*.

***

In this book I have often referred to the essence of the human self and that of all beings. The self-emergence of 'Nature' needs beholding, in awe

---

259.EHP p.152.

and wonder, by the poetic soul so that beings may appear in their unique *essence*. The imagery of the threshold attempts to regain that experience. What does it mean for beings to appear in their essence? Of course, essence usually indicates a fundamental characteristic that appropriately and necessarily belongs to something. Although this is not incorrect, we have sensed that Heidegger takes it further to insist that it is *that* to which something genuinely belongs, rather that it being its 'property'. Contrary to the dominant world-view, it is in this manner that the earth places its claim upon us rather than the reverse.

Let me bring this into focus by way of a personal example.

I am a keen bushwalker and one morning decided to walk to a high hill (in Australia we like to call these 'mountains') located within a state forest. I climbed it about 2 years earlier when it was accessible via an unpretentious little track to be rewarded by excellent views from rocky outcrops. Despite being surrounded by evidence of much logging in recent years, it still had a sense of remoteness, of solitude and stillness. As I thought of the mountain and anticipated its venturing, a sense of its essence was already with me. However, upon arriving near the start of its ascent I was dismayed to find that the small track was now a 30metre wide bull-dozed 'fire-break'. The devastation is being taken to its completion by trail-bikers who now revel in its conversion to a 'thrills and spills' theme park. The view was still there, but that seemed to matter little. My sense of loss was overwhelming. This is more than a simple loss of beauty and a sense of despoliation of the natural environment. It certainly is all that, but it was also an awareness of a profound bereavement that remained with me. It was the fundamental nature of loss itself that touched me. The essence that drew me to the hill has been so violated that it seemed to have lost its 'being'. The mountain had not only lost something that it previously 'possessed'; it had been torn from the essence to which *it* properly belonged. As a being it was no longer granted this essence; its homelessness now mirroring that of its despoilers. It was no longer able to disclose the essence that it previously 'brought to light'. The essence, whose features we recognise as remoteness, solitude, beauty, stillness and 'wildness', and whose beholding in awe and wonder drew me to the peak, had withdrawn. Therefore I could no longer regard it 'in innocence'; its impoverishment became my impoverishment. The gods had fled the threshold.

The above event also exemplifies the way in which language participates in this process of dispossession. Today our obsession with absolute control has been given the ultimate tools of satellite positioning systems and Google Earth. They enable us to observe the entire planet. Every square metre is now laid bare for our assessment, manipulation and unqualified management. So I 'Googled' the above peak and, sure enough, there it was with its new firebreak duly marked by the statutory body charged with it 'management'[260]. The accompanying assessment declared that it was "typical of the rocky outcrops in the area" and that the construction of the break would have insignificant effect.

The mountain is in no way unique; it has been *averaged out* as 'typical' and therefore can be dealt with as such. Not being interested in what it is in its essence means that not much is expected from it; it is *ordinary*. Words have the ability to not only convey meaning, but also to conceal it and numb us into acquiescence. This is how language, instead of *gathering* beings where they can glow in their essence, is also able to obliterate difference. For Heidegger difference entails not simply the features that differentiate things, the difference between two entities, but rather the event of consequence whereby Being gathers things in their essence. Difference then is by way of prior and essential unity and becomes an interplay of belonging and differing, and of uniqueness and resemblance. In this interplay beings appear in Being's light, whereby the features of their essence are perceived as ever more wondrous. The ordinary becomes extraordinary; aesthetics becomes more originary. Of course, there are still those who sense this emerging of nature and express the astonishing experience in diverse ways. Here is an ecologist describing the 'art' of nature that is in the everyday not taken to be of any real value because it is "not framed, labelled, critiqued and priced": "...the fabulous brush strokes on bark, exquisite cross-hatching on dragonfly wings, sinuous slug trails, sublime wavy leaves..."; a common mosquito seen magnified: "striking black and silver form, bold bands, starry spangles and a complex script of sweeping lines"[261]. Such vivid evocation of landscape cannot take place without human beings being 'open' to the essence to which it belongs.

The above seemingly insignificant incident is an example of the 'oblivion of Being', which can be repeated in countless ways, wherever the loss of 'the sacred' marks the destitution of our times. In Australia, as no doubt in other affluent countries, the shorelines are crusted with the homes of those prepared to expend the considerable money required in exchange for the 'possession' of sea views. With our backs to the astonishing diversity of life forms and the exquisiteness of the landscape we stare out to sea in homelessness. Many Australians have yet to truly arrive in this country as their focus is still elsewhere. As one aboriginal commentator remarked, "They live *on* Australia, not *in* it. They pass *over* the landscape, not *through* it"[262]. We call the countryside 'the bush' and therefore tend to only see bush: monotonous, with little apparent value, ready for the bulldozer. The response to the increasing threat of bushfires due to climate change is an escalating obsession with control; more clearing of 'fuel' (i.e. habitat), more 'prescribed burning', which often results in additional problems. It

---

260. A technical issue: much of the track is so steep (and therefore susceptible to extensive erosion) that fire fighting equipment would not be able to access it. Due to its location and topology its efficacy as a fire break is demonstrably imaginary. No doubt the status of the surrounding forests as "available resources" precipitated this futile effort at defending these possessions.

261. Tim Low, GMagazine (NSW), (7), 2008, p.85.

262. Neil Murray, writing in *The Age*, 'Was True Blue a blackfella?', July 6 2002.

sometimes seems that Australians have a total incomprehension of the land in which they live. What if we were to turn around and truly grasp the essence to which the landscape belongs? What seemed like untidiness might become a hint that there is a lot happening in the bush without our 'help'. If we are attentive we might become aware there is much emerging and withdrawing going on. We would no longer see drought, flood and fire as the adversary to be mastered at all cost. We would not attempt to abstract nature to suit our theoretical streamlined models which time and again are turn out to be mere 'boxed-in' caricatures that cannot grasp the 'naturalness' of nature, and therefore are doomed to failure. When we understand ourselves as being-in-nature the quality of our interactive conduct becomes consistent with the very environment itself in which these are conducted. While not failing to respond to the urgent challenge of the rising global effects of human arrogance on the biosphere, we might also understand that drought, flood and fire are natural, and ultimately uncontrollable, experiences of the Australian land. Perhaps if we surrender to its 'magic' we may see these seemingly unforgiving characteristics as the emerging fabric of its essence in which we are inexorably interwoven. We would learn to dwell in this fabric in a way that respects its being. Rather than imposing our conception of what nature *should* be like, we might celebrate the harshness of the landscape that illuminates the softness of the green lands from which many of us began life's journey.

To perceive beings in this light we must not stray too far from the pulse of the earth, and from the source that is Earth. Rather than clinging to nostalgic dreams of distant lands across the waters we need to remain near to the shore of the *Sea*, the inexhaustible text of Being itself; near to where we can taste the brine that enlivens and sharpens the creative spirit. Close enough to undergo the wind and the tumbling waves that shake the foundations of autonomous familiar abodes. Here we are near enough to the sphere where the gods dwell to hear their abyssal silence and their need for being. As time intersects with space, and truth with thought, we touch on the origin and stand in awe and wonder. Homecoming is a return to this experience. Then, 'more experienced', when the song is left behind, the dance will carry on in our daily lives. Like specks glowing in a shaft of sunlight, what was previously only 'consciousness' is now illuminated by the truth of Being.

The following excerpts from lyrics of 'popular' music, perhaps demonstrate that poetic events may occur (often unrecognised) in diverse domains. It expresses a poetic speaking that senses a homelessness and the subsequent need for a sojourn in the turbulence of an openness beyond the horizon of the everyday. Even simply the awareness of 'strangeness' means the journey on a pathway of wandering has begun. Such wandering, away from 'logical' linear thinking, is essential as it engages with a primordial sense of human existence. The following lines seem to speak of, firstly an originary sense of awe and wonder, which is stifled by mere feelings and by the willing of the ego. Then follows a silence, which comes when the willing of 'the first beginning' is exhausted and the seeking of Earth begins as a response by the self to an enduring yet 'voiceless call' from the Sea, the 'great text of Being'. In the absence of customary time and space 'the soul' becomes strange and follows a nameless stranger into apartness; there is a foregoing of false comforts; a strange homeliness - 'far-off' - yet near.

In these reflections on the open of the threshold we begin to know the (ungrounded) ground in which we dwell; where the soul, as the 'innermost heart' becomes the stranger, as it follows the strange light of Being that goes before, on the way towards Earth, and sees the world anew:

> Heaven holds a sense of wonder
>
> Possessing all this beauty
>
> yet hungry for more
>
> I can't help this longing
>
> Passion chokes the flower
>
> until she cries no more.
>
> The thundering waves, the pounding Sea is calling me home
>
> In this white wave I am sinking
>
> I am sinking in this silence
>
> you are silent.
>
> It was there, at the crossroads of time,
>
> and I wondered why.
>
> Strange how
>
> my heart beats
>
> to find myself on this shore.
>
> Strange how
>
> I still feel
>
> My loss of comfort gone before.
>
> ...there before me a shadow
>
> ...where no other can follow
>
> Close to home - feeling so far away [263].

I suggest that this is not *thinking* in the threshold; awareness as to the significance of what is happening here may be largely absent. It is not, for instance, the poeticising thoughtfulness of Heidegger's friend, the poet and French resistance fighter René Char, whose work looked deeply and painfully into the totality and the uncertainty of the experience of Being. He too understood that the immortal gods may be encountered in a 'between', close to the darkness of the oceanic site:

---

263. A very selective borrowing from Loreena McKennitt – The Old Ways, Delirium-*Silence*, Enya- *On your shore*.

Just as there are several different nights in the universe, there are several gods on the shores of day. However they are so dispersed that between breathing in and breathing out a lifetime passes.

The gods do not die or go into decline. They withdraw by means of imperious and cyclical movements akin to those of the ocean. One does not come upon them unless they are lying submerged in tide pools[264].

Yet, even in the preceding selection, the strangeness of the soul is recognised. Perhaps it is more like the primordial Greek response to the awe and wonder of existence and to the angst of the void. Feelings there seem to retain a central role. Heidegger could perhaps be accused of disparaging the bodily, the sensual, feelings and emotions. Certainly, he went at great pains to insist that his thinking is not to be interpreted 'psychologically'. I too have stressed my central aim of seeking human authenticity, a grasping of what it means to be and being faithful to the ground of Being. I too have insisted that 'good feelings' are not the goal of our venture; these become of little consequence in the threshold. Yet, as a real feature of being human, feelings are also not to be belittled, or even less, purged from thought lest they contaminate the 'purity' of our thinking.

When we encounter the light or the truth of Being in the threshold, we may well anticipate delight and even elation as the imagination overflows and unexpected creative responses come to pass. We may also feel dread in the sense of the sublime, the numinous that rests in the abyss of Being and recalls our non-being. Yet we do not resort to religious ideas which only serve to make us *feel* 'at home in the unhomely' instead of coming to terms with this essentially human way of being. Feelings are now grounded; their amelioration is not the 'purpose' of the venture of the threshold. Although a sense of worth and a profound awareness of the meaningfulness of existence should be the outcome of the thinking in this book, this is not another self-righteous prescription on 'how to be happy'. There is no shortage of self-help books whose promises, that 'one's life will be changed forever' and imbued with happiness, health and fulfilment, can be conveniently consumed in the shortest possible time. This book does not evade the reality that in life there will always be moments that seem too painful to bear. To be sure, there is much evidence of the positive psychological and physiological healing benefits of reflection, of involvement in the arts, poetry and thoughtful literature. However, here I have tried to show that this is because these hold the indispensable ingredients for the nourishment of the human soul. They give us the possibility to re-awaken the human essence and to be faithful to it.

This book is an invitation, a hint as 'where to look' if we are to dwell in an authentically grounded human way. Therefore, as we must look until we begin to see, it cannot be an instruction manual that unambiguously lays out the steps to be taken to find our authentic selves. It instead unveils a less comfortable pathway that imagines the overflowing of self-contained

---

264.René Char, op. cit., p.167.

boundaries of biological and sociological interpretations. As the pathway is not trivial, it is not always easy going; the treasure it approaches is not a means to an end, but is that which conveys the sacred in earthly being. The distant source that is approached *is* the treasure, but only if it is experienced *as* the source. Here, in a new realm of meaning, we sense and express the unthought, the 'abyssal' nature of what lies 'beyond' it, thereby protecting and preserving an essential experience that can only belong to the kind of beings we are.

Certainly, the 'thoughtful thinking' of the threshold, as it is 'touched' by the truth of Being, does not to cling to single ideas or one-track courses. Threshold thinking does not go with the first thing that grabs us, but with that which at first sight goes unnoticed. Today there seems to be much that goes unnoticed. We tend to be drawn towards that, which seems disclosed as familiar and assured, and as appearance. It is not a failure or imperfection on our part; it is just a way we belong to this occurrence of being. It affirms that Being enables thinking to become familiar and secure with things that are disclosed in an everyday manner; this is *one* manner of belonging to being. Yet, if it becomes the only way of being the soul becomes so impoverished and forgotten that we are no longer able to encounter it as the spiritual dimension of our deepest essence.

I believe that today we need to stand back and see where the human journey has taken us and what kind of wisdom might help to get us back on track. Perhaps the 'death of God' might herald a new beginning that resurrects the human soul. This book has attempted to not only give an account of a more poetic and imaginative way of being that lays the foundation for a truly sustainable interaction with Nature, but also to give this pathway an intellectual and existential grounding. This is a pathway that brings us nearer to our essence, which resides close to the foundation of the spiritual and the sacred. It does not preclude a passionate relationship with the world, but also gives it the fertile ground that recalls an originary human comportment. Why is it supposed that sacredness and 'the holy' can only be the realm of the supernatural? This book does not deny that the universe is a natural realm, governed by nature's laws. However, my hope is that it has given a hint that sacredness is a characteristic of being that can be encountered, quite naturally, in the naturalness of Nature, when its 'natural' truth comes *near* to the essence of human beings. Even the gods that then come to pass do not evade this naturalness; indeed they embody this essential character of the world. The retrieval of the gods is an experience of spirituality that is without dogma and ritual. It is one that does not find refuge in the "security of salvation in celestial blissfulness"[265], or comfort under the guiding hand of a mythical supreme being. Nor does it sacrifice its authentic freedom for the norms of tradition and culture. In the threshold we are in the crossing, where we find an openness to truth that, as Günter Figal writes, "holds the middle between the enhancement of a

---

265. EHP p.137.

life that *only trusts its own force*, and the turning (in faith) towards a transcending power beyond"[266]. How much more astounding, yes even *beyond* astounding, is the retrieval of the essence of things in fidelity to the abyssal ground, compared with the fabricated systems of religion, contrived to assure an illusionary certainty out of what is ultimately unexplainable.

Of course, there are also those who hold that for Christian and other traditions *weakness* rather than power and certainty is at the heart of its message. They reject the warranty of correctness central to the orthodox religions. They scorn the prominence of emotion and feelings, and the gospels of success and prosperity, e.g. as expounded by some Evangelical churches. They emphasise that their faith also holds an ambiguous tension between strength and weakness, between the now and the need to await, between doing and being, and between freedom and bondage. Theirs is a God that is less certain and is more difficult to define and represent. It is not a personal God that intervenes in the daily affairs of human beings. They reject a focus on the self that excludes the poor, the lonely and the dispossessed in a suffering world. Even more, they refuse the wretched death wish of apocalyptic beliefs. For them powerlessness can only be encountered from a position of weakness. Doubt, hope and despair must be experienced before we can presume to attend to those of others. Notions of God are surely overturned here; such a faith, as it is less assured and surrenders to a sense of love, of grace and beauty, is on a shifting shore. Perhaps some sense in the image of Jesus an essence of humanity which has frailty and vulnerability at its heart. Perhaps it represents a move that begins to venture towards the abyss that is beyond faith[267]. Indeed, the Jesus event may well a moment of the passing of the gods if its essential truth is not suffocated. Yet, it seems, ultimately theology cannot quite let go of a faith in 'a being' and remains within the constraints of a subjectivity that does not seize the strange freedom of the sublime offering I have described. The crossing of the threshold offers a bridge between the worlds of the phenomenal, where men and women build, and the noumenal, from where 'the heavenly ones' come to pass. It is not as though these worlds are on opposite banks, but, by bringing out the unity and truth of *both* domains, we are rather to simultaneously transform them. The gods, as they approach our dwelling place, come as 'guests'. Will we provide the clearing in which the wanderer will find the refuge that shelters gods and humans; the guest-house for a homecoming of the soul and the spirit?

"And springing from our intoxicated brow a higher meditation
Commences at once with ours, the flowering of the heavenly,
And to the open gaze the illuminating one will be open"[268].

266.CCP p.208.

267.Cf. Jean-Luc Nancy, *Dis-enclosure,* op. cit.

268.Hölderlin, 'Remembrance', in EHP, p.143.

Abbreviations for cited works

BNG   Being, Nothing and God (Seidel)

BQP   Basic Questions of Philosophy (Selected "Problems" of "Logic")(Heidegger)

BT   Being and Time (Heidegger)

CCP   Companion to Heidegger's Contributions to Philosophy (Scott)

CP   Contributions to Philosophy (From Enowning)(Heidegger)

CPI   Heidegger's Contributions to Philosophy – An Introduction (Vallega-Neu)

EHP   Elucidations of Hölderlin's Poetry (Heidegger)

EP   The End of Philosophy (Heidegger)

ET   The Essence of Truth (Heidegger)

FT   A Finite Thinking (Nancy)

ID   Identity and Difference (Heidegger)

IM   Introduction to Metaphysics (Heidegger)

M   Mindfulness (Heidegger)

OWL   On the Way to Language (Heidegger)

P   Parmenides (Heidegger)

PA   The Portable Atheist (Hitchens)

PLT   Poetry, Language, Thought (Heidegger)

QCT   The Question Concerning Technology and Other Essays (Heidegger)

SE   The Song of the Earth (Haar)

TAH   Thinking after Heidegger (Wood)

TB   On time and Being (Heidegger)

TWH   Thinking with Heidegger (de Beistegui)

WCT   What is called thinking? (Heidegger)

ZS   Zollikon Seminars (Heidegger)

Bibliography

Arendt, Hannah. *The Life of the Mind*, New York: Harcourt, Inc, 1978.

Badiou, Alain. *Being and Event*, New York: Continuum, 2005.

de Beistegui, Miguel. *Thinking with Heidegger – Displacements*, Bloomington: Indiana University Press, 2003.

Char, René. *This Smoke that Carried Us*, trans. Susanne Dubroff, New York: White Pine Press, 2004.

Derrida, Jacques. *The Gift of Death*, trans. David Wills, Chicago: University of Chicago Press, 1995.

Fisher, Frank. *Response Ability*, Elsternwick: Vista Publications, 2006

Foltz, Bruce V. *Inhabiting the Earth: Heidegger, Environmental Ethics, and the Metaphysics of Nature*, New Jersey: Humanities Press International, 1995.

Haar, Michel. *Heidegger and the Essence of Man*, trans. William McNeill, Albany: State University of New York Press, 1993.

------ *The Song of the Earth*, trans. Reginald Lilly, Indiana University Press: 1993.

Heidegger, Martin. *Basic Questions of Philosophy (Selected "Problems" of "Logic")*, trans. R. Rojcewicz and A. Schuwer, Bloomington: Indiana University Press, 1994.

―― *Being and Time*, trans. Joan Stambaugh, Albany: State University of New York Press, 1996.

―― *Contributions to Philosophy (From Enowning)*, trans. P. Emad and K. Maly, Bloomington: Indiana University Press, 1999.

―― *Discourse on Thinking*, trans. J. Anderson and E. Freund, New York: Harper & Row, 1966.

―― *Elucidations of Hölderlin's Poetry*, trans. Keith Hoeller, New York: Humanity Books, 2000.

―― *Identity and Difference*, trans. Joan Stambaugh, Chicago: University of Chicago Press Edition, 2002.

―― *Introduction to Metaphysics*, trans. Gregoryu Fried and Richard Polt, London: Yale University Press, 2000.

―― *Mindfulness*, trans. Parvis Emad & Thomas Kalary, London: Continuum, 2006.

―― *Off the Beaten Track (Holzwege)*, Cambridge: Cambridge University Press, 2002.

―― *On the Way to Language*, trans. P. D. Hertz & J. Stambaugh, New York: Harper & Row, 1982.

―― *On Time and Being*, trans. Joan Stambaugh, Chicago: University of Chicago Press Edition, 2002.

―― *Parmenides*, trans. Adré Schuwer & Richard Rojcewicz, Bloomington: Indiana University Press, 1992.

―― *Pathmarks*, ed. William McNeill, Cambridge: Cambridge University Press, 1998.

―― *Poetry, Language, Thought*, trans. Albert Hofstadter, New York: HarperCollins, 2001.

―― *The End of Philosophy*, trans. Joan Stambaugh, Chicago: University of Chicago Press Edition, 2003..

―― *The Essence of Truth*, trans. Ted Sadler, London: Continuum, 2002.

―― *The Fundamental Concepts of Metaphysics: World, Finitude, Solitude*, trans. William McNeill & Nicholas Walker, Bloomington & Indianapolis: Indiana University Press, 1995.

―― *The Metaphysical Foundations of Logic*, trans. Michael Heim, Bloomington: Indiana University Press, 1984.

―― *The Question Concerning Technology and Other Essays*, trans. William Lovitt, New York: Harper & Row, 1977.

―― *Towards the Definition of Philosophy*, London: The Athlone Press, 2000.

―― *What is called thinking?*, trans. J. Glenn Gray, New York: Harper & Row, 1968.

―― *Zollikon Seminars*, ed. Medard Boss, Illinois: Northwestern University Press, 2001.

Hitchens, Christopher (ed.). *The Portable Atheist*, US: Da Capo Press, 2007.

Kisiel, Theodore. *The Genesis of Heidegger's Being &Time*, Berkeley: University of California Press, 1995.

Leunig, Michael. *The Lot – In Words*, Penguin, 2008.

Lovitt, William and Harriet Brundage. *Modern Technology in the Heideggerian Perspective*, Vol. I & II, UK: The Edwin Mellen Press, 1995.

Mathews, Freya. *'Journey to the source of Merri Creek'*, Melbourne: Ginninderra Press, 2003.

Merleau-Ponty. M. *Phenomenology of Perception*, London: Routledge &Kegan Paul, 1962.

Nancy, Jean-Luc. *A Finite Thinking*, Stanford: Stanford University Press, 2003.

------, *Dis-enclosure – The Deconstruction of Christianity*, New York: Fordham University Press, 2008.

Pöggeler, Otto. *Martin Heidegger's Path of Thinking*, New York: Humanity Books, 1991.

Rojcewicz, Richard: *The Gods and Technology –A reading of Heidegger*, Albany: State University of New York Press, 2006.

Scott, Charles E. *et al* (eds). *Companion to Heidegger's Contributions to Philosophy*, Bloomington: Indiana University Press, 2001.

Seidel, George J. *Being, Nothing and God*, Assen: Koninklijke Van Gorcum & Comp., 1970.

Thiele, Leslie Paul. *Timely Meditations – Martin Heidegger and Postmodern Politics*, Princeton: Princeton University press, 1995.

Thomson, Iain. 'The Philosophical Fugue: Understanding the Structure and Goal of Heidegger's *Beitrage*', *Journal of the British Society for Phenomenology*, 34(1), Jan.2003, pp.57-73.

Vallega-Neu, Daniela, *Heidegger's Contributions to Philosophy – An Introduction*, Bloomington: Indiana University Press, 2003.

Wood, David. *Thinking after Heidegger*, Oxford: Blackwell Publishers, 2002.

Young, Julian. 'Death and Transfiguration: Kant, Schopenhauer and Heidegger on the Sublime', *Inquiry*, vol.48, No.2, 131-144, April 2005.

CPSIA information can be obtained
at www.ICGtesting.com
Printed in the USA
LVOW13s1318120318

569542LV00012B/334/P

9 781863 356312